The Latin American Studies Book Series

Series Editors

Eustógio W. Correia Dantas, Departamento de Geografia, Centro de Ciências, Universidade Federal do Ceará, Fortaleza, Ceará, Brazil

Jorge Rabassa, Laboratorio de Geomorfología y Cuaternario, CADIC-CONICET, Ushuaia, Tierra del Fuego, Argentina

Andrew Sluyter, Louisiana State University, Baton Rouge, LA, USA

Maria Elisa Belfiori · Mariano Javier Rabassa
Editors

The Economics of Climate Change in Argentina

 Springer

Editors
Maria Elisa Belfiori
School of Business
Universidad Torcuato Di Tella
Buenos Aires, Argentina

Mariano Javier Rabassa
Departamento de Investigación Francisco
Valsecchi
Facultad de Ciencias Económicas
Universidad Católica Argentina
Buenos Aires, Argentina

ISSN 2366-3421   ISSN 2366-343X (electronic)
The Latin American Studies Book Series
ISBN 978-3-030-62254-1   ISBN 978-3-030-62252-7 (eBook)
https://doi.org/10.1007/978-3-030-62252-7

This Springer imprint is published by the registered company Springer Nature Switzerland AG
The registered company address is: Gewerbestrasse 11, 6330 Cham, Switzerland

Foreword

Argentina presided the G20 Summit in 2018 inspired by the motto "Building Consensus for Fair and Sustainable Development." The fact that attaining fair and sustainable growth was the motto of the summit evidences the birth of a new kind of agenda. This new agenda includes everybody, the scientific and academic community and, especially, the civil society who has been demanding sustainable development for many years without success. It is not the current government's credit because our positions last only a definite time but the result of a cultural change that must transcend.

From this perspective, the creation of the National Cabinet for Climate Change, under Executive Branch Decree 891/2016, was an important institutional achievement. The Paris Agreement in December 2015 was also a turning point for Argentina, as it now compels our country to take responsibilities. To ratify the goal of a 15% reduction of greenhouse gas emissions by 2030 compared to the projected business as usual emissions, our government had to carry out a reorganization process. As an Executive Branch Decree created the National Cabinet for Climate Change, it is the President who must call the ministries to participate and, through the Chief of the Cabinet of Ministers, must coordinate the agenda. Then, the ministers must establish focal points to carry it out, and the Secretary of Environment and Sustainable Development coordinates and facilitates the work. The first step in the agenda was to validate the national contribution introduced at COP Marrakesh in November 2016. For this, the National Cabinet for Climate Change called in the ministries with higher impact on global warming and the provinces. There were also roundtables held with experts and members of the civil society. Nevertheless, the goal will eventually be attained only if we can design sectorial plans and the relevant players validate them. In November 2017, three sectorial plans were presented at the COP in Bonn transforming the national goal in concrete actions: energy, transportation, and forests.

In this search for concrete actions consistent with equitable and sustainable development, the government is not alone. The scientific community, universities and civil society also have a role to play according to their incumbencies. Generally, public policies underestimate the value of consulting with the scientific and academic community. Our government understands, however, that we need not only goodwill and good intentions but also the solidity and consistency of the scientific work to achieve good results. The scientific community and universities must encourage logic, and not ideological, discussions and foster debates in a democratic, diverse, and free environment. It is tough to develop a public policy without a baseline to measure. If you cannot measure it, you cannot evaluate it; if you cannot evaluate it with available and measurable data, it is not a public policy. In this sense, the major deficit of our country as regards the environmental issues has to do with statistics and baseline measurement. Therefore, to rearrange the area, we need the strategic contribution of science and universities.

In line with these management policies, the Secretary of Environment and Sustainable Development is working with two concrete actions as baselines: a national inventory of landfills and a reforestation plan. There are three reasons why we should choose national inventories of landfills as a measurement of the anthropic activities generating an irreversible damage to the citizens' quality of life. First of all, there are many people still living in dumpsites, which is unacceptable. In a circular and powerful economy undergoing a paradigm change, garbage should be a waste turning into a resource. Therefore, if we fail to measure the base, we will not change this situation easily. Second, as we measure garbage, we are also measuring quality of life, as well as air, water, and earth purification. Finally, our children and future generations are nowadays conscious that the culture of recycling, reducing, and reusing is a must.

As regards the reforestation plan, it is aimed not only at achieving "zero deforestation," but also at curing the damage that has already been done. We should differentiate guilt and responsibility here. I am not guilty, but I feel responsible. We can neither stop climate change nor promote actions to take responsibility for what we have done without a proactive policy of adding green to what has been degraded. That is why the reforestation plan, as our management indicator, has both technical and ethical components. The ethical component makes us accountable for the future and, in an exemplary way, seeks to remedy and repair the damage we have done.

Together with these elements that are the core of our environmental policy, there is an autarchic entity called Administración de Parques Nacionales (National Park Administration). It used to be part of the Secretary of Tourism, but it is now within the structure of the Secretary of Environment and Sustainable Development. It represents an essential asset for Argentina in terms of environment and sustainable development. Protected areas and parks are an asset that we must preserve and protect at all costs, without regression or irreversibility. Accordingly, the President decided to double the surface of protected and preserved areas as a sustainable development goal. And the most critical innovation within that goal is ocean preservation and the creation of the first two marine protected areas and national sea parks. This includes Yaganes Park and Banco Burwood II, as well as the Marine Protected Area Namuncurá. Notice that, in general, when we are asked to draw the map of Argentina, we outline the continent, but take the sea for granted. Even though we always include the Malvinas Islands and the Argentine Antarctica, we tend to forget that there is also a rich territory in our seas. In this way, we will preserve 10% of marine areas and 300 km^2

of marine protected areas, expanding horizons and strategically including the sea in our preservation policies.

Many of these environmental and sustainable development policies demand processes and times that exceed any administration's term of office. Said policies will remain as a firm basis to be carried out and continued in the future. Consequently, a tension appears between medium- and long-term policies, and short-term needs. Medium- and long-term policies demand vision. Short-term policies, however, frequently require the fulfillment of needs that cannot always be satisfied with structural reforms. Even more, medium- and long-term policies need strategies and a vision that frequently have an impact in terms of short-term electoral policies. Electoral policies are necessary in a democracy, but they are not necessarily good in terms of generating long-term policies that exceed any administration term of office. From this perspective, the support of the scientific community and universities must be sustained in the long term.

Speaking of the environment office, our most challenging task is transversality. We have to work across different ministries, jurisdictions, and disciplines. The government of each province is responsible for their environmental issues because they are autonomous and sovereign states. They may use and take possession of natural resources and, at the same time, control them. They are autonomous in terms of the anthropic activities they carry out within their territories. At the same time, the National Government is responsible for social and environmental minimum requirements that establish a minimum standard, but it is not entitled to intervene. Should the government of a province not comply with said minimum requirements, the National Government can only resort to starting legal action against the province before a federal court. When citizens realize that the provinces are not observing the minimum environmental requirements, they demand the National Government enforces the law. This is a legitimate right of people, but it is also one of the countercultural effects of change. Human beings, in general, and Argentine people, in particular, always expect "someone else" to comply with the law. They say: "I'll do it, but not before someone else, or the government, does it. I won't be the first one." However, even though the government makes decisions, a cultural change must accompany these decisions. The "social responsibility agenda" must be transversal. It requires not doing the same things a bit better, but doing them differently. It requires a paradigm change, that is to say, a cultural shift that takes longer than one administration's term of office.

Concerning this environmental agenda, the Pope's Laudato Si Encyclical introduces a concept that is mandatory for all states, from an ethical, theological, and also secular point of view. It appeals to all people without distinctions, to all religions and to non-believers to take care of our common home. If there is something that may unite all human beings under a shared cause, it is precisely taking care of our common home. A discussion on national and jurisdictional boundaries is no longer meaningful, as the impact of global warming and its effects spread beyond borders. From a political and economic view, the most polluting countries should contribute with more resources to mitigate and adapt to climate change damages. It is not a matter of developed or underdeveloped countries, but of justice and level of responsibility. If we measure temperature increase from the time of the Industrial Revolution up to now, why should we not identify those who generated said increase from that time onward?

Fighting global warming must be not only an aspiration but also an obligation. This means that we must leave great ideas and attitudes aside and speak of money. Therefore, beyond any discussion of practical philosophy and metaphysics, what we cannot put in monetary terms will not have any impact. What we cannot explain in figures and economics will be a noble intention by people of goodwill, aspiring to protect the planet. I am not objecting to this vocational attitude, but it belongs to the civil society and nonprofit organizations. As regards public policy, what is not profitable is ineffective. In other words, so long as we do not turn profitable into sustainable, the search for sustainable development will continue to be just a testimonial attitude. Therefore, economic and environmental issues are the same thing. Governments need academic and scientific arguments to make the market rules change. However, if we talk about market rules, we also have to talk about incentives: rewards and punishments. Rewards and punishments not only in terms of "you pollute, you pay," but also in that nobody will be willing to carry out a business that does not fit the sustainability criteria, because that business will not be cost-effective.

Non-tariff barriers to trade also play a role. Climate policies will end up as a non-tariff barrier to trade. Therefore, we must adapt to the sustainability criteria; otherwise, we will be out of the international market. We want to remain in this market, but under our own parameters. We are not willing to accept any restrictions by third parties based on models that do not fit us. Our scientists and experts must participate in the global discussions with models that fit our economy. We do not want, for example, to measure our cows' emissions based on the criterion adopted in Europe, because our production system is different. Instead of taking external standards or values, we want to apply our own resources: our people. We also want a serious discussion. To attain this, Argentina must be consistent and predictable, and our position has to be validated by data. And all this has to be done without changing the order, as changing the order may, in this case, change the product. Decisions must be based on previous academic and scientific measurements and evaluations, and not political preconceptions. This may lead us to uncertain scenarios because open, honest, and sincere discussions based on data generate turbulence. Nonetheless, this discussion is welcome, as it leaves aside non-validated judgments and prejudgments. This is why our government has high expectations that universities will contribute with forums and discussions to consolidate the parameters on which policy discussions should be based.

Buenos Aires, Argentina   Sergio Bergman
                                    Minister of the Environment and Sustainable
                                    Development (December 2015–September 2018)

Preface

This book covers topics on the economics of climate change, mostly from a Latin American perspective. It is the product of papers' presentations and discussions that took place at the "First Workshop on Environmental Economics and Energy" held at The Pontifical Catholic University of Argentina, in the City of Buenos Aires, in April 2018.

Climate change is the most pressing problem and the biggest challenge that humanity is facing nowadays. Countries around the globe are seeking alternative strategies to mitigate and adapt to the consequences of global warming. Latin American countries' role in shaping the regional climate agenda is yet to be explored. This book seeks to contribute to this policy agenda by putting together the research carried out by Argentinian economists on issues related to climate change, natural resources, the environment, and the economy.

In this volume, economists investigate the impacts of climate change and the policies to mitigate and adapt to the global temperature increase from an empirical and theoretical perspective. The potential health impact of climate change is at the forefront of the concerns. In the first chapter, García-Witulski and Rabassa explore the effect of extreme temperatures on mortality risks, emphasizing gender and age differences.

In the second chapter, Ahumada and Cornejo analyze the effects of global warming on agricultural yields, using data of soybean crops in Argentina. Agricultural production is highly sensitive to climate change. A considerable decline in crop production can result from global temperature increases and changing rainfall patterns. This production drop would be a threat to the fulfillment of global needs for food, animal feed, and biofuel.

In the second part of the book, contributors offer a macroeconomic perspective on climate change. Conte Grand asserts that the only way to meet the Paris Agreement goals with sustained growth is by decoupling economic activities from carbon emissions. The chapter explores the potential conflict between growth and emission reduction based on the international experience of different countries from 1990 to 2012.

Efforts to establish a global climate policy have been unsuccessful, despite years of climate negotiations. Thus, governments are exploring alternative strategies to mitigate and adapt to global warming's consequences within their jurisdictions. Ramos and Chisari evaluate the costs and benefits of a unilateral carbon tariff in Argentina. The authors conclude that it is an expensive policy for developing economies that usually face persistent unemployment, inequality, and external and fiscal imbalances, and volatile capital flows.

In the last chapter of this part, Belfiori provides a general view of the essential elements of optimal climate policy. The author questions the role of renewable energy subsidies and explains some paradoxes connected to the political tendency to subsidize renewables instead of taxing carbon.

Finally, Carlos Winograd closes the book with a discussion about the key challenges that Latin America will face in the coming decades. The chapter examines the prospects of a development path based on the region's response to an increase in world food demand. The author analyzes the environmental and social tensions that can arise from such a development path.

This compilation of articles is intended for academics, researchers, policy-makers, and a general audience interested in the economics of climate change and the steps that countries can take to move forward. It has been written not only for social scientists but also, and especially, for people around the world interested in climate change impacts, adaptation strategies, and climate policy.

Younger generations are showing higher awareness of the climate problem. They are the hope that the world is getting closer to solving this critical issue. This book is written for them, too, trusting that it will bring to the table the tools and analytical framework that they need to support their views and visions.

Buenos Aires, Argentina   Maria Elisa Belfiori
July 2020   Mariano Javier Rabassa

Acknowledgements

The book *The Economics of Climate Change in Argentina: Impact Studies and Mitigation Policies* is the output of the papers' presentations and discussions that took place at the "First Workshop on Environmental Economics and Energy" held in Buenos Aires during April 2018 and hosted by The Pontifical Catholic University of Argentina. The support of The Pontifical Catholic University of Argentina in organizing and hosting this workshop throughout the years is gratefully acknowledged.

Contents

Contributors

Eduardo Andrés Agosta Facultad de Ciencias Astronómicas y Geofísicas, Universidad Nacional de La Plata, La Plata, Argentina;
Consejo Nacional de Investigaciones Científicas y Técnicas, Buenos Aires, Argentina

Hildegart Ahumada Universidad Torcuato Di Tella, Buenos Aires, Argentina

Maria Elisa Belfiori School of Business, Universidad Torcuato Di Tella, Buenos Aires, Argentina

Omar Osvaldo Chisari CONICET-Universidad de Buenos Aires, Instituto Interdisciplinario de Economía Política (IIEP-Baires), Buenos Aires, Argentina

Mariana Conte Grand Department of Economics, Universidad del CEMA, Ciudad Autónoma de Buenos Aires, Argentina

Magdalena Cornejo Universidad Torcuato Di Tella and CONICET, Buenos Aires, Argentina

Lucio Florio Facultad de Filosofía y Letras, Universidad Católica Argentina, Buenos Aires, Argentina

Christian Martín García-Witulski Facultad de Ciencias Económicas, Pontificia Universidad Católica Argentina, Ciudad Autónoma de Buenos Aires, Argentina

Mariano Javier Rabassa Facultad de Ciencias Económicas, Pontificia Universidad Católica Argentina, Ciudad Autónoma de Buenos Aires, Argentina

María Priscila Ramos Departamento de Economía, Facultad de Ciencias Económicas, Universidad de Buenos Aires, Buenos Aires, Argentina;
CONICET-Universidad de Buenos Aires, Instituto Interdisciplinario de Economía Política (IIEP-Baires), Buenos Aires, Argentina;
Centre d'Études Prospectives et d'Information Internationale (CEPII), Paris, France

Mariana Paula Torrero Facultad de Ingeniería y Ciencias Agrarias, Universidad Católica Argentina, Buenos Aires, Argentina

Carlos Winograd Paris School of Economics and University of Paris-Evry Val
d'Essone, Paris, France

Climate Change: Impacts and Adaptation Policies

The Impact of Temperature on Mortality in Argentinean Municipalities

Christian Martín García-Witulski and Mariano Javier Rabassa

Abstract Although earlier studies have documented the influence of weather on human health in major Argentine cities, these studies lack national coverage. Moreover, the use of different health metrics and weather parameters makes it impossible to assess any geographic heterogeneity of impacts. We aimed to analyze the effect of hot and cold temperatures on human mortality rates in Argentina and their differential impacts by age and sex. The study rests on nonparametric techniques applied to data with a panel structure to estimate the causal effect of temperature extremes on mortality risks. Our findings confirm that extreme temperatures increase mortality rates relative to mean monthly temperatures, but the impact of colder than average temperatures is larger in magnitude. On average, a day with a countrywide mean temperature of 5 °C increases overall monthly mortality rate by 3.5% points with respect to the observed mortality for a day with countrywide average temperature. As expected, there exists substantial heterogeneity between age groups, with older people facing larger risks. To a lesser extent, there are also heterogeneous impacts by gender and geographic regions. These findings provide relevant information for policy makers about potential impacts of changing temperatures over Argentina in the upcoming decades.

Keywords Temperature · Climate change · Mortality · Vulnerability · Argentina

Introduction

Climate change represents a major threat for humans and other forms of life in our planet. The potential impact of climate and climate variability on the social and environmental determinants of health outcomes is at the forefront of scientific

C. M. García-Witulski (✉) · M. J. Rabassa
Facultad de Ciencias Económicas, Pontificia Universidad Católica Argentina, Avenida Alicia Moreau de Justo 1400, C1107AAZ Ciudad Autónoma de Buenos Aires, Argentina
e-mail: christian_garcia@uca.edu.ar

M. J. Rabassa
e-mail: mariano_rabassa@uca.edu.ar

© Springer Nature Switzerland AG 2021
M. E. Belfiori and M. J. Rabassa, (eds.) *The Economics of Climate Change in Argentina*, The Latin American Studies Book Series,
https://doi.org/10.1007/978-3-030-62252-7_1

and policy concerns. Recently, the World Health Organization (WHO) reported that climate change is expected to cause approximately 250,000 additional deaths per year worldwide between 2030 and 2050, due to malnutrition, diarrhea, and heat stress (World Meteorological Organization 2018).

Presently, the global mean temperature is approximately 1.2 °C above the pre-industrial era (World Meteorological Organization 2017). There exists a strong agreement among the scientific community regarding the human influence on the observed and predicted changes in temperature and precipitation patterns. For instances, the Intergovernmental Panel on Climate Change (IPCC) states that anthropogenic influences have very likely contributed to the observed rise in terrestrial and ocean temperatures since 1970. Without further attempts to curb these emissions, it is predicted that the global mean temperature will rise between 3.7 and 4.8 °C by the end of the century, increasing extremely hot days while decreasing cold ones (Intergovernmental Panel on Climate Change 2014). Understanding the effects that extreme temperatures pose on human health is key to promote effective adaptation policies to rising temperatures from climatic change.

Potential impacts of weather changes on mortality are usually divided between direct impacts, i.e., those produced by temperature increases or weather-related natural disasters, and indirect impacts arising mainly from changes in vector-borne diseases and food security. This paper concerns with the former: the direct effect that extreme temperatures have on mortality risks.

The relationship between thermal extremes and mortality has been extensively documented in the literature for developed countries (Baccini et al. 2008; Basu and Samet 2002; Basu et al. 2005; Braga et al. 2002; Curriero et al. 2002; Medina-Ramón and Schwartz 2007) and to a lesser extent in developing countries. In Latin America, the evidence is pretty thin (Bel et al. 2008; Romero-Lankao et al. 2013). In general, these studies find excess mortality during both cold and hot periods, although this relationship varies greatly by geographic regions. Also, they report differences in population susceptibility, with women and the elderly at most risk. For a more detailed survey of findings, see Basu and Samet (2002) and Basu (2009).

A few studies have documented the influence of weather on human health in major Argentine cities. The closers to our work are Almeira et al. (2016), which reports a U-shaped relationship between temperature and mortality for the cities of Buenos Aires and Rosario, and De Garín and Bejarán (2003), which examines the effect of thermal stress during summertime, characterized by the relative strain index, on mortality rates in Buenos Aires city. Other study, using data from emergency room visits at a hospital in central Buenos Aires reports the association between weather conditions and several pathologies for wintertime and summertime, during the 1996–97 season (Rusticucci et al. 2002). Yet other studies have turned their focus into specific pathologies, particularly respiratory diseases. For instance, Piccolo et al. (1988) documented the relationship between asthma hospitalizations and meteorological variables in Bahia Blanca. More recently, two studies for the city of Córdoba find that a higher daily mean temperature and a wider daily temperature range are important determinants of infectious diseases in both the upper and lower respiratory tracts, particularly among the elder population, and in low-income households facing housing deprivation (Amarillo and Carreras 2012; Carreras et al. 2015).

Although very informative, these studies lack national coverage. Moreover, the use of different health metrics and weather parameters makes it impossible to assess any geographic heterogeneity of impacts. In a context of climatic change, where resource to promote adaptation to the new climate norms must be allocated, comparable countrywide estimates of health costs are most needed to develop a rational adaptation policy.

This paper makes two contributions to the literature. First, it extends on a very thin literature on climatic determinants of mortality in Argentina. Second, it provides the first countrywide and comparable regional estimates of the incidence of temperature on mortality, breaking it up by gender and age group.

Econometric Modeling

The causal effect of temperature on mortality rates in Argentinean municipalities is analyzed using nonparametric models for panel data, i.e., repeated cross-sectional data. Nonparametric models estimated by ordinary least squares have been widely applied in the epidemiological literature (Barreca and Shimshack 2012; Curriero et al. 2002; Doyon et al. 2008; Kaiser et al. 2007) because the relationship between weather and mortality rates is likely to be nonlinear (Basu 2009). Thus, the estimated benchmark model is described by the following equation:

$$\text{TM}_{cpmy} = f\left(\text{temp}_{cpmy}\right) + f\left(\text{prep}_{cpmy}\right) + \mu_m + \omega_{cm} + \epsilon_{cpmy} \qquad (1)$$

where TM_{cpmy} is the mortality rate per 100,000 inhabitants in municipality c within province p in month m in year y. The final unit of analysis is a "municipality by month." The exposure functions $f\left(\text{temp}_{cpmy}\right)$ and $f\left(\text{prep}_{cpmy}\right)$ are additive functions containing parameters to be estimated related to exposure to monthly average temperature and precipitation, respectively. μ_m stands for time effects that allow controlling for potentially confounding effects common to all municipalities that vary overtime, for example technological advances or macroeconomic shocks affecting the countrywide economic performance, which might affect health outcomes. Finally, ω_{cm} is an interaction term which captures temporal variations between municipalities due to differences in air quality, income levels, educational attainment, and population density. Thus, these fixed effects could be interpreted as a baseline estimate of mortality rates in each municipality (Barreca and Shimshack 2012).

The exposure functions are estimated by means of restricted piecewise cubic splines, using the mkspline package in the statistical software Stata, version 14.0. Restricted cubic spline performs better than a linear spline when working with very curved functions (Rosenberg et al. 2003), thus allowing both mortality and weather to vary more flexibly without needing to choose a priori any functional form. A total of six knots were chosen for both the average monthly temperatures. The number of knots seems appropriate for large datasets, and the specific values were selected using the percentile distribution (Curriero et al. 2002; Rosenberg et al. 2003) of the weather variables (Harrell 2001).

The estimated exposure parameters cannot be directly interpreted; therefore, they should be compared to a reference value, which in this case was selected at 18.9 degree Celsius (°C), the national mean daily temperature. It can be calculated as (Stata 2017):

$$V_{i+1} = \frac{(V_1 - k_i)_+^3 - (k_n - k_{n-1})^{-1} \left\{ (V_1 - k_{n-1})_+^3 (k_n - k_i) - (V_1 - k_n)_+^3 (k_{n-1} - k_i) \right\}}{\left(k_n - k_1^2 \right)} \quad (2)$$

where k_i is the value for the ith knot ($i = 1, \ldots, n$), V_i is the exposure parameter to be estimated ($i = 1, \ldots, n - 1$), and V_1 is the reference temperature (18.9 °C). Finally, $(u_+) = u$ if u is strictly positive, and $(u_+) = 0$ if u is zero or negative.

Data

The estimation strategy outlined above relies on two major sets of data between the years 2004 and 2010: death counts by administrative subdivision and weather records from meteorological stations.

The mortality data corresponds to the universe of 1,812,745 million deaths occurred across the country, compiled by Dirección de Estadísticas e Información Pública, which is the agency in charge of the national system of health statistics. The dataset reports each death along with the date of death, the gender and age, and the municipality of residence of the deceased. Thus, death counts had to be aggregated at the municipal level, the smallest administrative subdivision in the dataset.

Death counts were complemented with annual population estimates in order to construct all-cause monthly municipal mortality rates per 100,000 inhabitants, by gender and age groups. Annual municipal population had to be interpolated from the 2001 and 2010 national population censuses. Specifically, following Deschênes and Greenstone (2011) and Barreca et al. (2016), population was stratified in four age groups (0–4, 5–44, 45–64, and >64 years old), according to the population summaries in the 2001 census. Unfortunately, population summaries did not allow constructing separate mortality rates for infants, i.e., younger than one year old, which are far more susceptible to thermal stress than other young children (Keim et al. 2002).

Weather records were obtained from Servicio Meteorológico Nacional, the official weather and climatological agency. Daily maximum and minimum temperatures in degree Celsius, and total precipitation in millimeters (mm) are reported for 71 weather stations that comprise the national weather station network. Ten weather stations, all located in Patagonia, southern Argentina, had to be discarded due to incomplete weather records. The monitoring network might seem to be sparse given the surface area and the degree of development for a country such as Argentina. However, according to the 2010 national census, Argentinian population is mostly urban, with 91% living in towns bigger than 2000 inhabitants, and half of the total population living in just 8 cities. Therefore, these 61 weather stations included in the analysis provide coverage to approximately 82% of the country's population.

Daily station data had to be spatially interpolated at the municipal level to construct monthly weather exposure variables. Following a common approach in the literature, square inverse distance weights were calculated from each municipality geographical centroid to each station within a 100 km radius (Hanigan et al. 2006). Thus, the interpolated weather variables are simply the weighted average of the station records within that radius. Limiting the cutoff distance 100 km was necessary to balance the number of municipalities included in the study and minimizing the potential measurement error introduced by using distant stations. Similar cutoff levels have been used in empirical estimates for the USA (Barreca 2012; Barreca and Shimshack 2012; Barreca et al. 2016; Deschênes and Greenstone 2011; Ranson 2014).

Impact of Extreme Temperatures on Mortality

Panel A of Table 1 presents summary statistics of interpolated daily mean, maximum, and minimum temperatures for the 322 municipalities included in the study, stratified by census region, over the 2004–2010 period. It can be observed that the lower temperature records correspond to municipalities in Cuyo and Pampean regions—recall Patagonia has been excluded due to lack of complete weather station records. Minimum temperatures over these regions are on average 5.9 and 5.2 °C lower than in the warmer northeast region. As expected, municipalities in northern regions experience higher maximum temperatures, with records between 3.9 and 4.9 °C higher than the region with the lowest maximum temperatures.

Summary statistics for monthly mortality rates between age groups are reported in Panel B and by region in Panel C of the same table. Average mortality rates are 10.8 per 100,000 inhabitants for children under 5 year of age, 6.7 for people between 5 and 44 years of age, 47 for adults between 44 and 65, and 325 for older adults. Consequently, one should expect to observe larger impacts of extreme temperatures among these two last groups, given that they appear to the most vulnerable people. Looking at different regions, higher mortality seems to be correlated with higher daily mean temperatures, although one should be cautious to infer causation since confounding factors, for instance income levels, could be also correlated with climate. Notwithstanding, the identification strategy described in the previous section allows controlling for these confounding factors by including a series of fixed effects.

Figure 1 presents the estimated exposure impacts, and their respective 95% confidence intervals, computed by nonparametric regressions between mortality rates and temperature at the municipal level, controlling for precipitation levels, and fixed effects by month and municipality by month. The results must be interpreted as the relative impact of temperature (daily mean, maximum, or minimum) over annual mortality in percentage points (pp) due to an additional day in a given month with

Table 1 Descriptive statistics

Panel A. *Historical daily temperature data* °C (2004–2010, *n* = 25,409)

By region	Buenos aires metropolitan area	Pampean	Northwest	Northeast	Cuyo
(n %)	(4.5)	(44.04)	(25.69)	(13)	(12.74)
Average					
Mean	17.66	16.95	19.95	21.84	17.68
S.D.	4.79	5.21	5.55	4.48	6.22
Minimum					
Mean	12.59	10.72	13.08	16.01	10.27
S.D.	4.61	5.15	6.28	4.47	6.35
Maximum					
Mean	22.75	23.19	26.82	27.67	25.09
S.D.	5.03	5.47	5.14	4.67	6.29

Panel B. *Month mortality rate per* 100,000 *population by age* (2004–2010, *n* = 25,409)

By age	0–4	5–44	45–64	>64	–
(n %)	(100)	(100)	(100)	(100)	–
Mean	11.19	7.16	50.24	347.81	–
S.D.	25.96	10.22	45.58	206.68	–

Panel C. *Month mortality rate per* 100,000 *population by region* (2004–2010, *n* = 25,409)

By region	Buenos aires metropolitan area	Pampean	Northwest	Northeast	Cuyo
(n %)	(4.5)	(44.04)	(25.69)	(13)	(12.74)
Mean	59.28	61.62	31.88	39.77	44.21
S.D.	27.81	28.64	24.57	20.66	47.69

Notes The number of observations (*n* = 25,409) corresponds to municipality by month, i.e., 322 municipalities by 84 month, from January 2004 to December 2010. Observations with weather record with less than 25 days in a month were not included in the analysis (6% of total observations)

temperature above or below the observed average temperature during the 2004–2010 period.

For example, Panel A of Fig. 1 shows that on average, an additional day with a countrywide mean temperature of 5 °C increases overall annual mortality rate by 3.5 pp (95% CI: 3.06–3.96) with respect to the observed mortality for a day with countrywide average temperature (i.e., 18.9 °C). In general, Figure 1 shows a nonlinear relationship between temperature (daily mean, maximum, and minimum, respectively) and the overall mortality rate, i.e., all age groups, genders, and regions. A striking regularity depicted in Fig. 1 is that the estimated impacts of colder than average temperatures are higher than the computed estimates for hotter than average temperatures. Again, using Panel A to illustrate, any day with a mean daily temperature below the average, 18.8 °C, positively and significantly affects mortality rates,

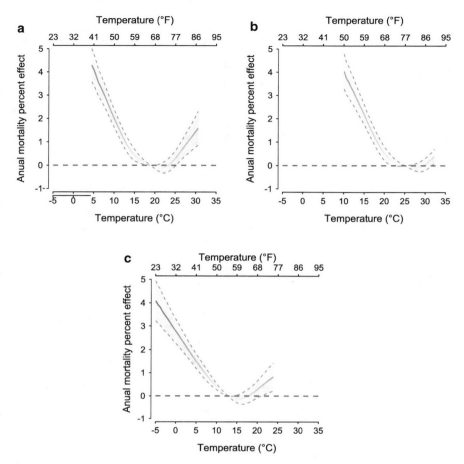

Fig. 1 Effect of temperature on overall mortality rate. The solid lines are the impacts of temperature on the mortality rate compared to a month with the average countrywide temperature. Blue lines are colder than average and red lines hotter than average temperature. Dashed line is the 95% confidence intervals. **A** Daily mean temperature (reference value is 18.9 °C). **B** Maximum daily temperature (reference value is 24.4 °C). **C** Minimum daily temperature (reference value is 12.2 °C)

and the estimated impacts increase as the temperature decreases. However, only those days with temperatures above 25.5 °C generate statistically significant impacts on mortality risks. Another interesting result is that while colder temperatures are always significant, only the highest maximum (Panel B) and highest minimum (Panel C) temperatures produce slightly statistical significant impacts on mortality.

Physiological factors and adaptation to the environment may affect the relationship between mortality risks and temperature. In order to examine potential heterogeneity in the temperature–mortality relationship, regressions were stratified by

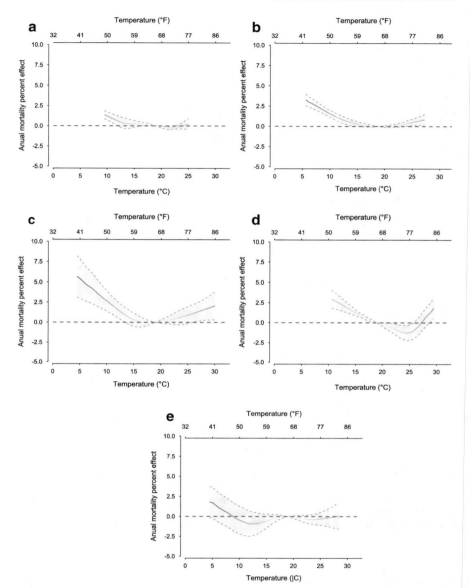

Fig. 2 Effect of daily mean temperature by census regions. The solid lines are the relative risks of mortality compared to a day with the average countrywide daily mean temperature (18.9 °C). Blue lines are colder than average and red lines hotter than average temperature. Dashed line is the 95% confidence intervals. **a** Buenos Aires metropolitan area. **b** Pampean. **c** Northwest. **d** Northeast. **e** Cuyo

census regions (Fig. 2), and age group and gender (Fig. 3). Figure 2 shows that for all regions, colder than (the national) average temperatures have at same point of the curve a positive and statistically significant impact on mortality rates. For instance, one additional day in a month with a mean daily temperature of 9.4 °C increases annual mortality by 4.6 pp (95% CI: 3.38–5.79) in Buenos Aires metropolitan area (Panel A). In the same fashion, an additional day of 5.6 °C rises annual mortality risks by 5.2 (95% CI: 2.37–8.08) in the Pampean region (Panel B). In the northeast (Panel C), a day of 8.33 °C increases mortality by 8.3 pp (95% CI: 0.54–16.05), while in the northwest region (Panel D) a day with mean daily temperature of 10.55 °C raises mortality by 7.83 pp (95% CI: 3.81–11.87). Finally, in the western region of Cuyo (Panel E) an additional day of temperature around 2 °C raises local mortality rates by 10.02 pp (95% CI: 2.47–17.57). Again, the estimated impacts are relative to a one less day in a month with mean daily temperature of 18.9 °C, which is the national average temperature.

Figure 3 provides evidence of heterogeneous impacts by age group (females in Panel A and males in Panel B). Only the elderly are affected by extreme temperatures, but colder than average temperatures seem to be more likely to be statistically significantly different from a day with 18.9 °C. Also, cold temperatures have a greater impact on male mortality risks, with no apparent difference between sexes for hotter than normal temperatures.

Discussion

Environmental factors are important determinants of human health. Recently, due to global warming and the recurrence of heat waves, the attention has been placed on how weather affects health outcomes, among which mortality has captured the most attention.

The existing literature, mainly for developed countries, has pointed to a U-shaped type of relationship between temperature and mortality risks. Using flexible regression models over longitudinal monthly weather and mortality data for 322 municipalities spanning January 2004 and December 2010, this study has confirmed this relationship from a countrywide perspective.

We found a significant impact of both cold and hot days on mortality risks; however, colder than average temperatures have a greater effect on mortality. This finding holds irrespective of using the daily mean, maximum, or minimum temperatures. Using a comparable methodology applied for US counties, Barreca and Shimshack (2012) also reported a similar association between colder temperatures and influenza mortality, although their work concerned mainly with the impact of absolute humidity. Almeira et al. (2016) also report a higher susceptibility to colder extremes than to hotter temperatures in the cities of Buenos Aires and Rosario. An additional interesting result is that hotter than average temperatures have no impact when splitting the data by regions. In particular, we found no significant impact of hotter than average temperatures in Buenos Aires metropolitan area. In contrasts, De Garín and Bejarán

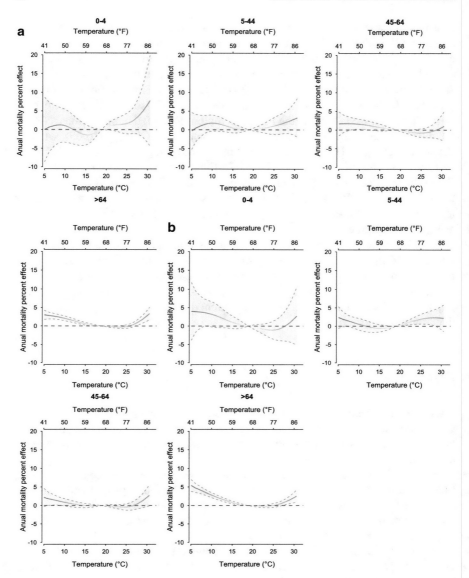

Fig. 3 Effect of daily mean temperature by age group and gender. The solid lines are the relative risks of mortality compared to a day with the average countrywide daily mean temperature (18.9 °C). Blue lines are colder than average and red lines hotter than average temperature. Dashed line is the 95% confidence intervals. **a** Females. **b** Males

(2003) reported that higher temperatures in Buenos Aires city, through its impact on the relative strain index, explain almost 10% of summertime mortality.

Our study has several limitations. First, the available population data did not allow constructing separate mortality rates for infants, i.e., younger than one year old, which are far more susceptible to thermal stress than other young children, as previously mentioned. Therefore, our results of no statistical significant influence of extreme temperatures on children's mortality rates have to be taken with caution. The second limitation is that mortality is obviously an extreme outcome, and thus underestimates the overall effect of extreme temperatures on human health. Unfortunately, there is no detailed countrywide medical database available to estimate both mortality and morbidity impacts. Another possible source of underestimation comes from the adoption of precautionary measures, particularly air conditioning, which have a protective effect during extreme temperatures (Barreca et al. 2016).

Another limitation concerns the use of weather data from meteorological stations. Given the sparse distribution of the station network, interpolation of weather parameters from distant stations might have introduced measurement error. To limit the chance of this type of error, we follow the convention of excluding weather station further than 100 km away from each municipal geographical center. Satellites might provide better weather information, at least in terms of spatial coverage (Kloog et al. 2014), but we leave this to future research. Also, we did not have access to information about mortality by cause. Therefore, the impacts of temperature on the overall mortality rate should understate the true impacts.

Having pointed out the potential short comes of this research, we believe this study provides valuable information on weather determinants of mortality in Argentina. The estimation approach employed in this papers allows to disentangle the causal effect of temperatures on mortality risk while controlling for possible confounders, reducing the possibility that the observed temperature–mortality link was driven by omitted factors associated with both weather and mortality.

To conclude, two main contributions of the paper are worth mentioning. Firstly, we produce the first countrywide and geographical comparable estimates of the temperature–mortality relationship in a sizable portion of the population. Second, we use repeated cross-sectional data instead of the more traditional time series approach used in previous empirical papers. This approach might help overcome the problem of relatively short time series usually observed in Argentinian data.

References

Almeira G, Rusticucci M, Suaya M (2016) Relationship between mortality and extreme temperatures in Buenos Aires and Rosario. Meteorologica 41(2):65–79

Amarillo AC, Carreras HA (2012) The effect of airborne particles and weather conditions on pediatric respiratory infections in Cordoba, Argentina. Environ Pollut 170:217–221

Baccini M, Biggeri A, Accetta G, Kosatsky T, Katsouyanni K, Analitis A et al (2008) Heat effects on mortality in 15 European cities. Epidemiology 19(5):711–719

Barreca A (2012) Climate change, humidity, and mortality in the United States. J Environ Econ Manage 63(1):19–34

Barreca AI, Shimshack JP (2012) Absolute humidity, temperature, and influenza mortality: 30 years of county-level evidence from the United States. Am J Epidemiol 176(7):114–122

Barreca A, Clay K, Deschênes O, Greenstone M, Shapiro JS (2016) Adapting to climate change: the remarkable decline in the US temperature-mortality relationship over the 20th century. J Polit Econ 124(1):105–159

Basu R (2009) High ambient temperature and mortality: a review of epidemiologic studies from 2001 to 2008. A global access science source. Environ Health 8(1):40

Basu R, Samet JM (2002a) An exposure assessment study of ambient heat exposure in an elderly population in Baltimore, Maryland. Environ Health Perspect 110(12):1219–1224

Basu R, Samet JM (2002b) Relation between elevated ambient temperature and mortality, a review of the epidemiologic evidence. Epidemiol Rev 24(2):190–202

Basu R, Dominici F, Samet JM (2005) Temperature and mortality among the elderly in the United States: a comparison of epidemiologic methods. Epidemiology 16(1):58–66

Bel IML, O'Neill MS, Ranjit N, Borja-Aburto VH, Cifuentes L, Gouveia NC (2008) Vulnerability to heat-related mortality in Latin America: a case-crossover study in Sao Paulo, Brazil, Santiago, Chile and Mexico City, Mexico. Int J Epidemiol 37(4):796–804

Braga ALF, Zanobetti A, Schwartz J (2002) The effect of weather on respiratory and cardiovascular deaths in 12 U.S. cities. Environ Health Perspect 110(9):859–863

Carreras H, Zanobetti A, Koutrakis P (2015) Effect of daily temperature range on respiratory health in Argentina and its modification by impaired socio-economic conditions and PM10 exposures. Environ Pollut 206:175–182

Curriero FC, Heiner KS, Samet JM, Zeger SL, Strug L, Patz JA (2002) Temperature and mortality in 11 cities of the eastern United States. Am J Epidemiol 155(1):80–87

De Garín A, Bejarán R (2003) Mortality rate and relative strain index in Buenos Aires city. Int J Biometeorol 48(1):31–36

Deschênes O, Greenstone M (2011) Climate change, mortality, and adaptation: evidence from annual fluctuations in weather in the US. Am Econ J Appl Econ 3(4):152–185

Doyon B, Bélanger D, Gosselin P (2008) The potential impact of climate change on annual and seasonal mortality for three cities in Québec, Canada. Int J Health Geogr 7(23):1–12

Hanigan I, Hall G, Dear KBG (2006) A comparison of methods for calculating population exposure estimates of daily weather for health research. Int J Health Geogr 13(5):1–16

Harrell F (2001) Regression modeling strategies: with applications to linear models, logistic regression, and survival analysis. In: Springer series in statistics

Intergovernmental Panel on Climate Change (2014) Climate change 2014 synthesis report. Contribution of working groups I, II and III to the fifth assessment report of the intergovernmental panel on climate change. http://www.ipcc.ch/pdf/assessment-report/ar5/syr/AR5_SYR_FINAL_SPM.pdf. Cited 9 Aug 2019

Kaiser R, Le Tertre A, Schwartz J, Gotway CA, Daley WR, Rubin CH (2007) The effect of the 1995 heat wave in Chicago on all-cause and cause-specific mortality. Am J Pub Health 97(Suppl 1):S158:62

Keim SM, Guisto JA, Sullivan JB (2002) Environmental thermal stress. Ann Agric Environ Med 9(1):1–15

Kloog I, Nordio F, Coull BA, Schwartz J (2014) Predicting spatiotemporal mean air temperature using MODIS satellite surface temperature measurements across the Northeastern USA. Rem Sens Environ 150:132–139

Medina-Ramón M, Schwartz J (2007) Temperature, temperature extremes, and mortality: a study of acclimatization and effect modification in 50 United States cities. Occup Environ Med 64(12):827–833

Piccolo MC, Perillo GM, Ramon CG, DiDio V (1988) Outbreaks of asthma attacks and meteorologic parameters in Bahia Blanca, Argentina. Ann Allergy 60(2015):107–110

Ranson M (2014) Crime, weather, and climate change. J Environ Econ Manage 67(3):274–302

Romero-Lankao P, Qin H, Borbor-Cordova M (2013) Exploration of health risks related to air pollution and temperature in three Latin American cities. Soc Sci Med 83:110–118

Rosenberg PS, Katki H, Swanson CA, Brown LM, Wacholder S, Hoover RN (2003) Quantifying epidemiologic risk factors using non-parametric regression: model selection remains the greatest challenge. Stat Med 22(21):3369–3381

Rusticucci M, Bettolli ML, Harris De Los Angeles, M (2002) Association between weather conditions and the number of patients at the emergency room in an Argentine hospital. Int J Biometeorol 46(1):42–51

Stata Corp L (2017) Stata base reference manual release, vol 15. Stata Press

World Health Organization (2018) Climate change and health. http://www.who.int/news-room/fact-sheets/detail/climate-change-and-health. Cited 9 Aug 2019

World Meteorological Organization (2017) WMO statement on the state of the global climate in 2017. http://library.wmo.int/opac/index.php?lvl=notice_display&id=20220#.WsSZlI4zMt8. Cited 9 Aug 2019

How Econometrics Can Help Us Understand the Effects of Climate Change on Crop Yields: The Case of Soybeans

Hildegart Ahumada and Magdalena Cornejo

Abstract Climate econometrics is a new field, which is proving to be a fruitful approach to give a rigorous treatment to many hypotheses related to climate change. This chapter illustrates how econometrics can help understand the effects of climate change on the time behavior of crop yields at a country-level scale. We use a multivariate framework to discuss issues related to non-stationarity of climate variables, the existence of nonlinearities, and collinearities. We also examine the exogeneity of the explanatory variables in crop yields models, the presence of extreme events, and the short- and long-run effects of climate change. Global climate change may affect the incorporation of new lands to production and the rise of crop yields to meet the increasing demand for food and energy. However, adaptation, trade, the declining share over time of agriculture in the economy, and carbon fertilization have reduced the harmful impacts of global warming. In particular, carbon fertilization refers to the positive effects that increasing carbon dioxide concentrations have on some plants that use CO_2 during photosynthesis. We focus on soybeans in Brazil, USA, and particularly in Argentina, as interesting examples of mitigation and adaptation to global and regional climate fluctuations.

Keywords Climate change · Carbon dioxide · Crop yields · Adaptation · Mitigation

H. Ahumada
Universidad Torcuato Di Tella, Av. Figueroa Alcorta 7350, Buenos Aires, Argentina
e-mail: hahumada@utdt.edu

M. Cornejo (✉)
Universidad Torcuato Di Tella and CONICET, Av. Figueroa Alcorta 7350, Buenos Aires, Argentina
e-mail: mcornejo@utdt.edu

© Springer Nature Switzerland AG 2021
M. E. Belfiori and M. J. Rabassa, (eds.) *The Economics of Climate Change in Argentina*, The Latin American Studies Book Series,
https://doi.org/10.1007/978-3-030-62252-7_2

Introduction

Worldwide population growth is expected to increase by over a third—or 2.2 billion people—by 2050 according to the 2017 United Nations World Population Prospects revision. This projection poses an urgent need to increase agricultural production to meet global needs for food, animal feed, and biofuel, by increasing the amount of agricultural land to grow crops, enhancing productivity on existing agricultural lands, or both. The main challenge would be how to meet the increasing demand for food while protecting our natural resources.

Although the effects of climate change may imply incorporating new areas devoted to different crops, the possibility of extending lands to production or increasing crop yields on existing lands may be threatened by global climate change. Crop production and yield are highly dependent on climate. Increases in temperature and changing patterns of rainfall associated with global and local climate changes may lead to a considerable decline in crop production. Also, extreme weather events such as droughts, heat or cold waves, and heavy rainfall leading to floods have increased since 1950 according to the *Fifth Assessment Report* of the Intergovernmental Panel on Climate Change (IPCC). An analysis of long-term and short-term weather events is needed in order to measure the effects of climate change on crop yields.

Agricultural production needs to adopt more efficient and sustainable production methods to lessen the negative effects of climate change and to better tailor policies seeking to promote sustainable growth in the agricultural sector.

Adaptation to climate variability and extreme events can help in reducing vulnerability to long-term climate change. However, the effects of climate change also need to be considered along with other evolving factors that affect agricultural production, such as changes in farming practices and technology as well as commodity or input prices. Quantifying these effects will provide important insights into how much to spend on mitigation and adaptation. Furthermore, understanding and estimating the effects of climate change will also help policy-makers to develop mitigation and adaptation strategies.

Therefore, the aim of this chapter is to discuss how econometrics can help understand the effects of climate change on the time behavior of crop yields at a country-level scale. Indeed, climate econometrics provides an approach to give a rigorous basis for many hypotheses related to climate change. In this line, an accurate econometric strategy to model the effects of climate change on crop yields should be able to deal with, at least:

1 the non-stationarity nature of climate variables;
2 exogeneity;
3 the existence of nonlinearities;
4 the presence of extreme events;
5 disentangling short- and long-run effects of climate change;
6 collinearities in a multivariate framework.

We focus on soybeans in the main producer and exporter countries: Brazil and USA, and particularly in Argentina, as an interesting case of mitigation and adaptation processes due to global and local climate changes.

The chapter is organized as follows. Section "The Effects of Climate Change on Crop Yields" reviews the empirical literature of the effects of climate change on crop yields. Section "The Case of Soybeans" describes the case of soybeans. Section "On the Econometric Modeling of Soybean Yields" discusses different empirical issues that should be accounted by an accurate econometric model. The last section concludes.

The Effects of Climate Change on Crop Yields

Different approaches have been followed to study the effects of climate change on crop yields, many of them based on agronomic analysis. However, results are not conclusive about the effects of climate change, mainly due to the adaptation and mitigation strategies in agriculture that have been also implemented to alleviate its potential negative effects.

The negative effects of climate change are mainly associated to extreme high temperatures which are found to be harmful for crop growth (e.g., Chen et al. 2013). Crop yield losses on the hottest days drive much of the effect of temperature (Schlenker and Roberts 2009). In fact, many recent studies found that changes in temperature are more important than changes in rainfall, at least at the national and regional levels (Reilly and Schimmelpfennig 2000; Schlenker and Lobell 2010). Furthermore, crops are more sensitive to extremely high temperatures during the phases of the plant growth cycle (Auffhammer et al. 2012; Welch et al. 2010). Temperature extremes can be critical for reducing yields, especially if they coincide with the flowering stage of the crop (Wheeler et al. 2000). Burke and Emerick (2016) examine the effect of long-term changes in climate variables on yields using county-level data in the USA. Their results indicate that the main crops in the USA—corn and soybeans—are significantly and negatively affected by long-term changes in extreme heat temperatures.

Nevertheless, there would be several factors that have reduced the harmful impacts of climate change: adaptation, trade, the declining share over time of agriculture in the economy, and carbon fertilization. As stated by Nordhaus (2013, p. 84) "one important mitigation factor for agriculture is carbon fertilization". The carbon (or CO_2) fertilization effect is the phenomenon by which the increase of carbon dioxide in the atmosphere increases the rate of photosynthesis in plants. That is, the largest amount of carbon dioxide (CO_2) in the atmosphere, that has resulted from rising anthropogenic emissions, may have positive effects on the plant's growth as they use carbon dioxide during photosynthesis. Carbon fertilization has a greater effect on plants with C_4 and C_3 photosynthesis systems (such as corn and soybeans, respectively), which can concentrate carbon dioxide onto reaction sites.

According to Nordhaus (2013), multiple field studies found that doubling atmospheric concentrations of CO_2 would increase yields of rice, wheat, and soybeans about 10–15%. For the Argentine case, Magrin et al. (2005), by using agronomic models, found that increases in yields corresponding to climate changes between 1930–60 and 1970–2000 were 38% in the case of soybeans. In a recent paper, Ahumada and Cornejo (2019) found that the median variations on CO_2 concentrations in the atmosphere could have increased soybeans yields about 14% during 1973–2015, all else equal. However, the carbon fertilization effect may not take place if other plant growth factors are severely limiting. Nutrient levels, soil moisture, water availability, and other conditions must also be met. Gray et al. (2016) found that the intensification of drought eliminates the potential benefits of elevated dioxide for soybean.

Empirical studies should not ignore or underestimate the effects of adaptation measures as means for diminishing the adverse effects of climate change. Several adaptation measures such as shifting planting dates, rotating crops, or developing new crop varieties have also been suggested and implemented for reducing the vulnerability from the potential negative impacts of climate change on crop yield and production (Cohn et al. 2016; Lobell et al. 2008).

Even if the focus is on studying the effects of climate change on crop yields, an econometric model should be developed within a multivariate framework. That is, other potential determinants of crop yields apart from climate (e.g., technological factors) should be also considered in the analysis.

The Case of Soybeans

In this section, we describe the case of soybeans as an interesting example of mitigation and adaptation processes to climate change. Crop production and yields are highly dependent on climate, and, in fact, global climate change may threaten the incorporation of new lands to production or the increase of crop yields on existing lands. Climate changes and technological advances have shifted the main worldwide production areas to nowadays warmer latitudes (to the north in the Southern hemisphere and to the south in the Northern hemisphere).

The demand for oilseeds, and particularly for soybeans, is derived primarily from the commercial utilization of its sub-products, high-protein soybean meal for animal feed, and soybean oil for edible and inedible uses. Non-traditional soybean uses such as bio-energy and bio-products are expected to increase rapidly and promise to boost prices due to increased global demand and higher value added. Moreover, not only most renewable energy sources have minimal contributions to global warming emissions, in contrast to fossil fuels, but they also provide an alternative to the eventually reserves depletion.

Over the last decades, Brazil and Argentina's combined total soybean production has been greater than that of the USA (the world's top producer). According to the 2016/17 *World Agricultural Supply and Demand Estimates* of the U.S. Department of

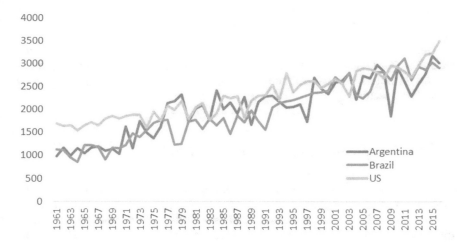

Fig. 1 Soybeans yield (in kg/ha) *Source* Own elaboration using data from FAOSTAT

Agriculture, Brazil and Argentina account for almost 49% of the worldwide soybean production compared to the 34% of the US production. Figure 1 shows the evolution of soybeans yield from 1961 to 2016 in Argentina, Brazil, and USA.

A rigorous empirical approach should be able to deal with the upward behavior of soybean yields and, thus, with their potential non-stationarity to identify which variables are the long-run determinants responsible of their observed trend.

Furthermore, new managerial practices have been introduced to increase crop yields and to adapt to climate change. For instance, no-till farming (also known as zero tillage) is an agricultural practice for growing crops without disturbing the soil through tillage. This practice decreases the amount of soil erosion by maintaining the soil structure and also protecting the soil by leaving crop residue on the surface. These practices have gained ground quickly in Argentina as an effective solution to the problem of soil erosion and the loss of nutrients, and, thereby, it increases carbon sequestration, reduces costs, and boosts productivity. The fastest adoption rates have been experienced in Argentina. According to AAPRESID, the Argentine No-till Farmers Association, 93% of the soybean area adopted the no-till system in the season 2016/17. The worldwide recognition of no-tillage farming as an effective sustainable system has spread no-till technology and other practices (e.g., crop rotation) to other areas and countries.

Other technological innovations include the use of genetically modified (GM) seeds. Soybean remains as the most adopted GM crop. According to the International Service for the Acquisition of Agri-biotech Applications (ISAAA)[1] biotech soybean accounted for 50% of all the biotech crop area in the world in 2016. Commercially

---

[1]ISAAA. 2016. Global Status of Commercialized Biotech/GM Crops: 2016. ISAAA Brief No. 52. ISAAA: Ithaca, New York.

grown GM soybeans are concentrated in a few countries, mainly USA, Brazil, and Argentina.

Therefore, given the technological advances experienced by this sector, empirical studies focused on studying the effect of climate change on crop yields should also control for mitigation and adaptation processes that are actually taking place in the agricultural sector.

On the Econometric Modeling of Soybean Yields

In this section, we discuss different empirical issues that an econometric model of soybean yield determination should consider in order to measure the effects of climate change in the agricultural sector. Hsiang (2016) offers a wide revision of the different econometric methods that can be used to study the effect of climate change on social and economic outcomes, in general.

Dealing with Trends

Variables may be classified according to their degree of time persistence into non-stationary or stationary. Non-stationarity is associated with the idea of long memory (high persistence) of past shocks on the behavior of a time series (e.g., crop yields). Such series could be stationary with a short-time dependence—that is, they could exhibit a significant tendency to mean reversion—after first differencing. In those cases, the series under study is said to have a unit root (a stochastic trend) or be integrated of first order, I(1).

This kind of behavior is generally compared with a typical model of deterministic trend to approximate the long-run behavior of a series. As stated by Lobell (2009), the trend in crop yields results largely from improvements in technology and, thus, for most crops the technology trend can be approximated with a first-order polynomial (a linear trend).

For the series of our interest, different unit root tests reported in Table 1 show that soybean yields can be represented as stationary around a deterministic linear trend. This trending behavior was also observed in Fig. 1. Because of that, many studies remove deterministic trends before studying the effects of climate factors on yields (see for example Thomasz et al. (2016) in the Argentine case or Tao et al. (2008) in the Chinese case).

Furthermore, given the nonlinear and non-stationary nature of crop yields, many detrending methods have been suggested to model them (see the comparison of detrending crop yield data techniques in Lu et al. 2017).

Nonetheless, it should be noted that if the aim of an empirical study is to understand which drivers could be behind the observed trending behavior, and the long-run relationships between crop yields and their potential determinants should be stud-

Table 1 Unit root tests, 1961–2016

Variable	Trend	k	ADF	b	PP	b	KPSS
ln yield^{ARG}	Yes	0	−5.49***	4	−5.77***	5	0.18**
ln yield^{BRA}	Yes	0	−5.53***	5	−5.45***	2	0.08
ln yield^{USA}	Yes	0	−7.66***	8	−8.03***	9	0.09
Δ ln yield^{ARG}	No	0	−14.29***	14	−23.52***	54	0.50**
Δ ln yield^{BRA}	No	2	−7.93***	23	−21.34***	20	0.20
Δ ln yield^{USA}	No	2	−7.48***	26	−30.83***	22	0.35*

Note k is the lag length selected by SIC, and b is the bandwidth using Bartlett kernel. *, ** and *** indicate significance at the 10%, 5%, and 1% level, respectively. ADF = Augmented Dickey Fuller, PP = Phillips-Perron, KPSS = Kwiatkowski-Phillips-Schmidt-Shin. The null hypothesis for the ADF and PP tests is that of a unit root against the alternative of stationarity. The KPSS test reverses the null and the alternative hypothesis. A constant and a trend were included for level variables, otherwise only a constant was considered

ied assuming them as I(1) (the simplest long memory process) and testing their co-integration (if the linear combination of those I(1) variables is stationary). Co-integration implies that two or more variables with a persistent behavior have common stochastic trends and that they will show a tendency to move together in the long run. As stated by Juseliu (2006, p. 18) "the order of integration of a variable is not in general a property of an economic variable but a convenient statistical approximation to distinguish between the short-run, medium-run and long-run variation in the data."

Moreover, a co-integration analysis could also be useful because it allows us to identify which variables move the equilibrium (the pushing forces) and which correct deviations from equilibrium (the pulling forces), that is, testing weak exogeneity as discussed in Sect. "Evaluating the Exogeneity of Climate Variables". We will also be back on the long-run and short-run effects on crop yields in Sect. "Disentangling Short and Long-Run Effects of Climate Change".

Evaluating the Exogeneity of Climate Variables

Typically, in the literature, climate variables (e.g., temperature, rainfall, humidity, storms, among others) are considered exogenous when studying their effect on agriculture, that is, they are considered as given for explaining crop yields. Due to the assumption about exogeneity and randomness of climate in many economic applications, climate variables act as a "natural experiment" and, therefore, would allow the researcher to statistically identify the causal effect of a variable on an economic outcome of interest.

However, as Pretis (2017) warns, human activity (say, through deforestation) affects local and global climate, and climate change, in turn, affects human activity (say, crop production or yields). Empirically, this implies that if we want to estimate

Table 2 Granger causality test, 1973–2015

Hypothesis	Statistic	p-value
ln yield does not GC temp	5.31	0.07
ln temp does not GC yield	9.20	0.01
ln yield does not GC CO_2	5.87	0.05
ln CO_2 does not GC yield	7.27	0.03

the effect of humanity (in its multiple dimensions) on climate change and vice versa, it is necessary to evaluate the exogeneity of the variables within the economic climate system to understand these interrelations in the long run. The analysis of exogeneity is crucial to obtain rigorous empirical estimates before proposing a model on economic variables or on climate variables. That is, the estimates from a single-equation model or a system model would be different depending on the variables exogeneity assumptions.

To estimate a model in which climate variables affect soybeans yields as expressed in Eq. (1), that is, climate variables as explanatory variables (also known as conditioning variables) in a single-equation model, exogeneity is a key assumption. However, once co-integration is found it is possible to evaluate exogeneity. A weak exogenous variable influences the long-run path of other variables in the system, and, at the same time, it is not influence by them. That is, the weak exogenous variable pushes to move the long-run relationship while the endogenous variable adjusts to maintain the equilibrium. To consistently estimate Eq. (1) as a single-equation model, we need to assume that yields are the endogenous variable while global temperature anomalies (temp) and CO_2 concentrations in the atmosphere (CO_2) are weakly exogenous.

$$\ln \text{yield} = \beta_0 + \beta_1 \text{temp} + \beta_2 \ln CO_2 + u \qquad (1)$$

Furthermore, if those climate variables are found to be weakly exogenous, we can test for non-Granger causality.[2] A variable will Granger-cause (GC) another variable if past values of a variable (say, global temperature or CO_2) contain information that helps predict another variable (say, soybeans yields). Thus, Granger causality is a statistical concept of causality based on the anticipation of variables.

Using data from 1973 to 2015, we estimated a climate system based on Argentine soybeans yields, global temperature anomalies, and global CO_2 concentrations in the atmosphere using two lags. Results, as shown in Table 2, indicate that both climate variables GC soybeans yields. With a significance level of 5%, the test rejects the null hypothesis that global temperature and CO_2 do not GC soybeans yields, but not vice versa. That is, on annual basis, climate variables (global temperature anomalies and CO_2 concentrations) anticipate soybeans yields. However, at a 10% significance level, results are not conclusive.

---

[2]It should be noted that Granger causality is a different statistical concept which is not a necessary condition for weak exogeneity, but for strong exogeneity (Engle et al. 1983).

However, as a long-run concept, weak exogeneity can give different results from those obtained analysis Granger causality. Ahumada and Cornejo (2019) found that all variables adjust to deviations from the long-run equilibria. This finding implies that climate variables are not weak exogenous which indicates that a system approach should be followed instead of estimating a single-equation model.

Although this result may be unexpected at first sight, it may be properly interpreted when we take into account the effect of deforestation. The soybeans upward trend in Argentina, a behavior also shown by other soybean producers such as Brazil (see Fig. 1), could have given incentives to the expansion of agriculture through the use of new lands coming from deforestation. Using data from NASA's Moderate Resolution Imaging Spectrometer (MODIS) on the Terra and Aqua satellites, Morton et al. (2006) have shown that in 2003, the peak year of deforestation in Matto Grosso (Brazilian state with the highest deforestation and soybean production rates) more than 20% of the state's forests were converted to cropland.

Deforestation contributes to global climate warming since it is responsible for not compensating the anthropogenic emissions of carbon dioxide to the atmosphere, and, in fact, deforestation releases CO_2 to the atmosphere. Therefore, from this perspective, we could think that global temperature anomalies and CO_2 emissions may also adjust to deviations from the long run in our estimated climate system.

Measuring Nonlinear Effects

Linearity is usually a starting functional form of many econometric models although there are several routes to relax this assumption by including polynomial terms, asymmetries, thresholds, etc. Different climate variables may have nonlinear relationships with crop yields.

Using different spatial panel econometric techniques, Chen et al. (2013) found nonlinearities and asymmetric relationships between yields and weather variables as it has been suggested in the literature. It is usually considered that the best predictor of crop yield is some measure of extreme heat during the growth period of the plant, considering a temperature threshold above 29 or 30 °C (Schlenker and Roberts 2009), depending on the analyzed crop. Furthermore, extreme high temperatures are harmful for crop growth, particularly during the phases of the growth cycle (Auffhammer et al. 2012; Welch et al. 2010). Therefore, the effect of temperature could be nonlinear but with a threshold at certain high levels.

Using daily data of maximum temperature from 54 meteorological stations of the Argentine soybean production area from 1973 to 2015, Ahumada and Cornejo (2019) constructed different variables that measure the number of days during the growing phase of the crop (from December to April) in which the temperature exceeded a threshold of 28, 29, 30, or 31 °C. They evaluated which of these different temperature thresholds has the most significant impact on Argentine soybean yields. The maximum temperature of each meteorological station was weighted by its share in the total soybean planted area. Those weights were annually updated to account for

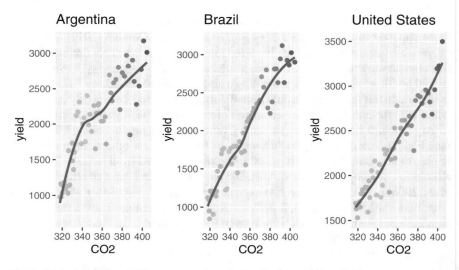

Fig. 2 Soybeans yield and CO_2 concentration relationship from 1961 to 2016

the displacement of crop areas over time. The construction of those variables allow to capture nonlinearities in the relationships between yield and local temperature as suggested in the literature. They found that ten additional days of maximum temperature above 30 °C during the growing season produce a decrease of 5% in soybean yields, all else equal.

For the Chinese case, Tao et al. (2008) also found a negative relation between growing season maximum temperature and soybean yield. Although they do not address the possibility of a nonlinear effect, soybean yield significantly decreased by 11.1–22.7% in the Shanxi province for each degree increase in growing season maximum temperature from 1979 to 2002.

Furthermore, as shown in Fig. 2, the relationship between soybean yields and CO_2 concentration in the atmosphere (to measure the carbon fertilization effect) may not be always linear, as in the Argentine case suggesting a second degree polynomial functional form. This is different to the observed linear behavior in USA and Brazil. Data on atmospheric CO_2 at Mauna Loa Observatory were obtained from the NOAA Earth System Research Laboratory.[3]

However, the figures and analysis were based on bivariate comparisons. In a multivariate framework, where other variables apart from yields and CO_2 are also considered, the inclusion of other controls may capture the possible nonlinearities and, consequently, a linear or log-linear functional form may be appropriate.

Many simulated crop yield responses studies evaluate crop yield response to elevated CO_2 assuming hyperbolic behavior. Long et al. (2006) review the CO_2 fertilization effects for the major C_3 and C_4 crops derived from enclosures, such as

---

[3]The carbon dioxide data on Mauna Loa constitute the longest record of direct measurements of CO_2 in the atmosphere.

controlled environmental chambers, transparent field enclosures, or open-top chambers. The fertilization factors averaged across the C_3 crops (rice, wheat, and soybeans) are 24% for yield.

Assessing the Effect of Extreme Events

According to the *Fifth Assessment Report* of the Intergovernmental Panel on Climate Change (IPCC), extreme weather events such as droughts, heat or cold waves, and heavy rainfall leading to floods have increased since 1950. In this sense, variables that measure extreme weather events are of interest to include in the econometric models.

El Niño and La Niña are opposite phases that occur when the interaction between the Pacific Ocean and the atmosphere above it changes from their neutral state in a cycle known as the El Niño-Southern Oscillation (ENSO). El Niño and La Niña events in the Pacific Ocean can be identified through the Southern Oscillation Index (SOI). The SOI, obtained from the Australian Bureau of Meteorology (BOM), is calculated using the pressure differences between Tahiti and Darwin. Sustained negatives values of SOI below -7 indicate El Niño episodes, while sustained positives values of SOI above $+7$ indicate La Niña episodes.

For the Argentine case, Ahumada and Cornejo (2019) found that an extreme event associated with La Niña episodes (droughts) decreases soybean yields between 1 and 2% during 1973–2015. However, no significant effects were found for El Niño events.

Tack and Ubilava (2013) estimate the impact of both El Niño and La Niña on average corn yields using a US county-level panel spanning 55 years (1950–2005). They estimate the impact of El Niño (La Niña) on mean corn yields as the percentage change in the mean of the El Niño (La Niña) regime relative to the mean of the neutral regime. The county-level impacts range from -24 to 33% for El Niño and from -25 to 36% for La Niña. The negative effect of El Niño is found for the corn belt region, and, as a result, the impact turns positive as one migrates both to the East and West.

The effects of El Niño and La Niña on crop yields may be accentuated if global warming leads to an increase in the frequency and intensity of this type of events.

As Nordhaus and Reinhard (2016) warn, extreme weather events are expected to increase worldwide and, given the local nature of some climate variables (particularly precipitations), the effects can also be measured at a local level. Using an unbalanced panel of 334 farms from 2002 to 2013, they analyze extreme weather events in the case of winter wheat throughout the Netherlands. They found that high-temperature events and precipitation events significantly decrease yields.

Disentangling Short- and Long-Run Effects of Climate Change

For series with persistent behavior, both stationary or integrated, which do not change much from period to period, it is possible to distinguish short-run and long-run effects.

For integrated variables, to obtain the long-run effects, we can test if the variables are co-integrated. Co-integration implies the existence of a linear combination of variables which is stationary when the variables in the relationship are non-stationary and integrated of equal order. Co-integrated variables are driven by the same persistent shocks, and, thus, those variables have a common stochastic trend, showing a tendency to move together in the long run. As indicated by (Juselius 2006), in multivariate co-integration analysis, all variables are represented as stochastic and a shock to one variable is transmitted to all other variables via the dynamics of the system until the system finds its new equilibrium position.

Once co-integration is found, an equilibrium correction model (ECM) can be estimated. This model encompasses differenced variables as well as the deviations from the long-run or co-integrated relationship for integrated variables as expressed in Eq. (3). There are several advantages of this formulation associated not only with avoiding multicollinearity typically present in time-series data but also allowing a more intuitive interpretation of the estimates disentangling short- and long-run effects.

Moreover, another advantage of using this approach is the invariance of the co-integration property to the extension of the information set (Juselius 2006). This property implies that once co-integration is found among a set of variables, the co-integration results will remain valid if more variables are added to the partial system, as the one we estimate below. In this sense, there would be no omitted variable effects present for co-integration when adopting this specific-to-general strategy.

Because of that, for the Argentine case, we start by estimating a partial system through a vector autoregressive (VAR)[4] model between 1973 and 2015 among soybean yields, global temperature anomalies, and CO_2 concentrations in the atmosphere. This climate system also controls for two variables[5]: La Niña events and the number of days with maximum temperatures above $30\,°C$ during the growing season of the plant. Both variables negatively affect soybeans yields.[6]

Therefore, there is evidence from the climate system that one long-run relationship exists in which all variables adjust to reach the long-run equilibrium. Equation(2) represents the long-run relationship when writting ln(yield) as the dependent variable.

---

[4]A VAR model is a multivariate stochastic process model that can be used to capture the linear interdependences among the variables analyzed in the system. It generalizes the time-series AR model.

[5]These variables, that we previously selected according to their statistical significance, were included unrestrictedly in the system, that is, outside the co-integration vector.

[6]The VAR model passes all diagnostic tests at traditional levels and included a linear trend in the long run since the variables can co-integrate but may have different deterministic trends. These results are not reported but can be obtained from the authors upon request.

Equation (3) shows the climate ECM derived from Eq. (2). Standard errors are reported in parentheses.

$$\widehat{\ln \text{yield}} = \underset{(0.75)}{-2.34}\text{temp} + \underset{(2.75)}{11.94}\ln CO_2 \tag{2}$$

$$\widehat{\Delta \ln \text{yield}} = \underset{(3.82)}{-15.97} - \underset{(0.06)}{0.27}[\ln \text{yield}_{t-2} - \underset{(2.75)}{11.94}\ln CO_{2,t-2} + \underset{(0.75)}{2.34}\text{temp}_{t-2}]$$

$$- \underset{(0.004)}{0.01}\,Nina_{t-1} - \underset{(0.002)}{0.008}\text{max}30_t - \underset{(0.09)}{0.61}\Delta \ln \text{yield}_{t-1} \tag{3}$$

$$R^2 = 0.72,\ \hat{\sigma} = 0.11$$

Equation (2) shows that both global temperature anomalies (temp) and CO_2 concentrations in the atmosphere are long-run determinants of Argentine soybean yields. To analyze the magnitude of estimated coefficients, we can note that in this sample period, if the temperature changes as its median value during the sample (0.06 °C) soybean yields will decrease about 14% in the long run, all else equal. However, as a possible mitigation effect due to that fertilization properties of CO_2, yields will increase near 6% as a consequence of the median percentage variations of CO_2 concentrations in the sample (0.47%). This last result is known as the CO_2 fertilization effect, as it has been previously described.

The estimated ECM indicates that 27% of the deviations from the long-run equilibrium is corrected in two years.

As regards the climate variables short-run effects, Eq. (1) shows that apart from an autoregressive behavior of soybean yields, there are negative effects of La Niña events—associated with droughts periods—and cumulated days of high temperature (above 30 °C). An extreme event associated with La Niña episodes decreases yields in 1%, while ten additional days of maximum temperatures above 30° during the growing season produce a decrease of 8%.

The estimated system in Eqs. (2) and (3) can be also used for prediction purposes. In this case, the constancy of the parameter estimates is a key issue. We can evaluate if parameters are unchanged by observing the recursive estimates of the coefficients. From an initial sample, the observations are added one by one until the last observation is included. Figure 3 shows that the coefficient estimates are inside the 95% confidence interval. Thus, for this simple model (from the partial system), parameter stability is not rejected.[7]

It should be noticed that we have also tested for other climate variables such as El Niño events and weather variables associated with excessive precipitations and floods,[8] but they were found statistically insignificant. The greater importance of changes in temperature over changes in rainfall on crop yields was also found by Reilly and Schimmelpfennig (2000) and Schlenker and Lobell (2010).

---

[7]Given the goodness of fit that has been obtained.

[8]Those variables include the variance coefficient, the rainfall gini index, the precipitation concentration index, and the cumulative precipitation during the growing phase of the plant.

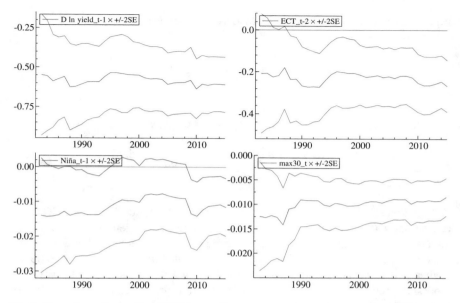

Fig. 3 Recursive estimates *Note* ECT denotes the error correction te rm, that is, the deviation from the long-run relationship

This partial system estimation and the ECM do not consider the effect of other relationships such as the interaction between technology and climate. As an example, using data from 1973 to 2015, we found on co-integration relationship between global temperature anomalies (temp) and two technological variables: the adoption of new seed varieties (seeds) and the no-till adoption (no-till). The adoption of new seed varieties is measured as the number of soybean varieties registered in USA, Brazil, and Argentina (the three main worldwide producers of soybeans). The no-till adoption is measured as the proportion of no-tilled cropland acres in Argentina.

In this co-integration relationship, which is reported in Eq. (4), only the adoption of new seed varieties adjusts to deviations from the long-run equilibria.

$$\widehat{\text{seeds}} = \underset{(0.74)}{2.13}\text{temp} + \underset{(0.005)}{0.01}\text{ no-till} \tag{4}$$

The positive long-run effect of global temperature anomalies on the use of new modified seeds may indicate that adaptation processes to climate change are being undertaken in the agricultural sector. According to these results, the potential negative effect of climate change on agricultural could have led producers to invest on technology in order increase their productions and crop yields. Moreover, the changes in managerial practices experienced in Argentina through the adoption of no-till practices has also a positive effect on the use of new seeds. The incorporation of no-till technology does not only protect the soil from erosion caused by plowing, but it also requires new genetically modified (GM) seeds (such as herbicide-tolerant crops).

Dealing with Collinearities in a Multivariate Framework

As it has been previously stated, in order to econometrically model the effect of climate change on crop yields, a multivariate framework should be used to control for and also to integrate different groups of determinants (climatic, technological, or economic) that may affect crop yields (see, for example, Huang and Khanna 2010). However, collinearity is often present in time-series data that show trending behavior. Moreover, Auffhammer et al. (2013) indicate that many empirical studies do not take into account all relevant climate dimensions, and many of them are difficult to measure. Therefore, estimating a model that includes different climate variables implies dealing with collinearities.

Some works focus on studying the effect of a single climate variable (such as temperature or rainfall). Auffhammer and Schlenker (2014) warn that if a single climate variable is used as a regressor, this measure will be subject to confounding variation of other climate measures that are correlated with it and also affect the variable of interest. This approach can lead to a classic problem of bias due to omitted variables.

In order to consistently estimate the long-run coefficient, a partial system approach can be followed as discussed in Juselius (2006). To take into account different potential determinants and to avoid collinearities, sub-systems of long-run relationships due to climate, technological, and economic factors can be estimated and then evaluated by encompassing of the different ECM representations (as in Ahumada and Cornejo 2019). The aim of testing encompassing is to address if crop yields adjust to one or several long-run relationships. We should recall the invariance of the co-integration property to the extension of the information set (Juselius 2006). This property implies that once co-integration is found in a partial system, the co-integration results will remain valid if more variables are added. Therefore, there would be no omitted variable effects present for co-integration when adopting this partial system approach.

Final Remarks

Effective adaptation strategies of crop production require long-run time-series analysis and a multivariate framework. In this line, the aim of this chapter has been to illustrate how econometrics can help understand the effects of climate change on the time behavior of crop yields at a country-level scale for the main producers and exporter of soybeans. We have focused on this crop as a particular case of adaptation and mitigation, with emphasis on the Argentine experience.

Climate econometrics provides an approach to give a rigorous basis for many hypotheses related to climate change. In this line, we have discussed different empirical issues that an accurate econometric strategy should address in order to model the effects of climate change on crop yields: the non-stationarity nature of climate variables; the exogeneity of the variables used for modeling crop yields; the existence of nonlinearities; the presence of extreme events; disentangling short- and long-run effects of climate change, and the problem of collinearities in a multivariate framework.

Using data from 1973 to 2015, the results obtained from a multivariate system estimation indicate that global temperature has a long-run negative effect of Argentine soybeans yields. Furthermore, we have also found negative short-run effects associated with La Niña events and high temperatures during the growing season of the plant. However, those global and local warming negative effects are partially mitigated through the CO_2 fertilization effect.

The estimation of this model has shown that a multivariate framework, including Niña events and the 30 °C threshold in temperature, and adopting a partial system instead of a single-equation approach give different results with a clearer interpretation of the estimates. It is worth noting that, apart from the long-run linear effect of global temperature anomalies, we found a short-run nonlinear effect derived from the number of days in which the maximum temperature exceeded 30 °C during the growing season of the plant.

So far, we have only focused on climate variables. However, when there are other potential determinants as in the case of crop yields, the econometric model should encompass other drivers like technology or economic factors. This appears as a future route of research in particular when technology should be analyzed as an adaptation response to climate change too.

Crop climate modeling may help identify which of the suggested adaptation strategies in the literature—the use of fertilizers, irrigation, change in planting date, no-till practices, among others—has significant effect in a specific location. Adaptation to climate variability and extreme events can help in reducing vulnerability to long-term climate change.

Quantifying these effects will provide important insights into how much to spend on mitigation and adaptation and thus, help policy-makers to develop those strategies.

References

Ahumada, H. and Cornejo, M. (2019), 'Are soybean yields getting a free ride from climate change? Evidence from Argentine time series'

Auffhammer M, Hsiang SM, Schlenker W, Sobel A (2013) Using weather data and climate model output in economic analyses of climate change. Review of Environmental Economics and Policy 7(2):181–198

Auffhammer M, Ramanathan V, Vincent JR (2012) Climate change, the monsoon, and rice yield in India. Climatic Change 111(2):411–424

Auffhammer M, Schlenker W (2014) Empirical studies on agricultural impacts and adaptation. Energy Economics 46:555–561

Burke M, Emerick K (2016) Adaptation to climate change: Evidence from US agriculture. American Economic Journal: Economic Policy 8(3):106–40

Chen, S., Chen, X. and Xu, J. (2013), 'Impacts of climate change on corn and soybean yields in China', *AAEA & CAES Joint Annual Meeting*

Cohn AS, VanWey LK, Spera SA, Mustard JF (2016) Cropping frequency and area response to climate variability can exceed yield response. Nature Climate Change 6(6):601

Engle, R. F., Hendry, D. F. and Richard, J.-F. (1983), 'Exogeneity', *Econometrica: Journal of Econometric Society* pp. 277–304

Gray SB, Dermody O, Klein SP, Locke AM, Mcgrath JM, Paul RE, Rosenthal DM, Ruiz-Vera UM, Siebers MH, Strellner R (2016) Intensifying drought eliminates the expected benefits of elevated carbon dioxide for soybean. Nature Plants 2(9):16132

Hsiang S (2016) Climate econometrics. Annual Review of Resource Economics 8:43–75

Huang, H. and Khanna, M. (2010), 'An econometric analysis of US crop yield and cropland acreage: implications for the impact of climate change', *AAEA Annual Meeting* pp. 25–27

Juselius K (2006) The Cointegrated VAR Model. Oxford University Press, Oxford, Methodology and Applications

Lobell DB (2009) Crop Response to Climate: Time-Series Models. In: Lobell DB, Burke M (eds) Climate change and food security: adapting agriculture to a warmer world. Springer Science & Business Media

Lobell DB, Burke MB, Tebaldi C, Mastrandrea MD, Falcon WP, Naylor RL (2008) Prioritizing climate change adaptation needs for food security in 2030. Science 319(5863):607–610

Long SP, Ainsworth EA, Leakey AD, Nösberger J, Ort DR (2006) Food for thought: lower-than-expected crop yield stimulation with rising co2 concentrations. Science 312(5782):1918–1921

Lu, J., Carbone, G. J. and Gao, P. (2017), 'Detrending crop yield data for spatial visualization of drought impacts in the united states, 1895–2014', *Agricultural and Forest Meteorology* pp. 196–208

Magrin GO, Travasso MI, Rodríguez GR (2005) Changes in climate and crop production during the 20th century in Argentina. Climatic Change 72(1–2):229–249

Morton DC, DeFries RS, Shimabukuro YE, Anderson LO, Arai E, del Bon Espirito-Santo F, Freitas R, Morisette J (2006) Cropland expansion changes deforestation dynamics in the southern Brazilian Amazon. Proceedings of the National Academy of Sciences 103(39):14637–14641

Nordhaus WD (2013) The climate casino: Risk, uncertainty, and economics for a warming world. Yale University Press

Nordhaus JP, Reinhard SJ (2016) Measuring the effects of extreme weather events on yields. Climatic Change 12(1):69–79

Pretis, F. (2017), 'Exogeneity in climate econometrics'

Reilly J, Schimmelpfennig D (2000) Irreversibility, uncertainty, and learning: portraits of adaptation to long-term climate change. Climatic Change 45(1):253–278

Schlenker W, Lobell DB (2010) Robust negative impacts of climate change on African agriculture. Environmental Research Letters 5(1):014010

Schlenker W, Roberts MJ (2009) Nonlinear temperature effects indicate severe damages to us crop yields under climate change. Proceedings of the National Academy of Sciences 106(37):15594–15598

Tack JB, Ubilava D (2013) The effect of El Niño Southern Oscillation on US corn production and downside risk. Climatic Change 121(4):689–700

Tao F, Masayuki Yokozawa JL, Zhang Z (2008) Climate-crop yield relationships at provincial scales in China and the impacts of recent climate trends. Climate Research 38(1):83–94

Thomasz, E. O., Casparri, M. T., Vilker, A. S., Rondinone, G. and Fusco, M. (2016), 'Medición económica de eventos climáticos extremos en el sector agrícola: el caso de la soja en Argentina', *Revista de Investigación en Modelos Financieros* 4

Welch JR, Vincent JR, Auffhammer M, Moya PF, Dobermann A, Dawe D (2010) Rice yields in tropical/subtropical Asia exhibit large but opposing sensitivities to minimum and maximum temperatures. Proceedings of the National Academy of Sciences 107(33):14562–14567

Wheeler TR, Craufurd PQ, Ellis RH, Porter JR, Prasad PV (2000) Temperature variability and the yield of annual crops. Agriculture, Ecosystems & Environment 82(1–3):159–167

Floods in Eastern Subtropical Argentina: The Contributing Roles of Climate Change and Socioeconomics

Eduardo Andrés Agosta, Mariano Javier Rabassa, Lucio Florio, and Mariana Paula Torrero

Abstract The roles of climate and socioeconomic components contributing to floods over eastern subtropical Argentina (ESA) region, the eastern portion of subtropical Argentina, are explored. The national structural flood risk map shows that the region manifests the national flood riskiest areas. Decadal mean flood risk values have peaked in the 1980s, though there is a positive trend in the subregion of Buenos Aires Metropolitan Area (GBA). One major natural factor to consider is flood hazard due to precipitation extreme events that have increased about 30% on average over the GBA subregion and northeastern ESA. The Paraná River shows positive streamflow trend attributable to climatic natural factors and climate change. Deforestation appears to be a high-pressure factor, due to decreased evapotranspiration and limited soil water storage capacity. Future climate projections indicate exacerbation of extreme events of precipitation and streamflow peaks. A critical aspect that makes the region further vulnerable is the lack of continuous and planned flood management policies. The increasing extreme rainfalls and flooding in the context of global climate change demand a complex approach of the phenomena.

E. A. Agosta
Facultad de Ciencias Astronómicas y Geofísicas, Universidad Nacional de La Plata, La Plata, Argentina
e-mail: eduardo.agosta@fcaglp.unlp.edu.ar

Consejo Nacional de Investigaciones Científicas y Técnicas, Buenos Aires, Argentina

M. J. Rabassa (✉)
Facultad de Ciencias Económicas, Universidad Católica Argentina, Alicia Moreau de Justo 1400, Edificio Santo Tomás Moro, C1107AFF Buenos Aires, Argentina
e-mail: mariano_rabassa@uca.edu.ar

L. Florio
Facultad de Filosofía y Letras, Universidad Católica Argentina, Buenos Aires, Argentina
e-mail: lflorio@uca.edu.ar

M. P. Torrero
Facultad de Ingeniería y Ciencias Agrarias, Universidad Católica Argentina, Buenos Aires, Argentina
e-mail: mariana_torrero@uca.edu.ar

© Springer Nature Switzerland AG 2021
M. E. Belfiori and M. J. Rabassa, (eds.) *The Economics of Climate Change in Argentina*, The Latin American Studies Book Series,
https://doi.org/10.1007/978-3-030-62252-7_3

35

Keywords Flood hazards · Vulnerability · Climate change · Subtropical Argentina

Introduction

Exposure to flooding is one of the most common natural hazards that global population faces. It is estimated that about 11% of people are currently living in flood-prone areas and 1% of the global population is, on average, exposed to floods each year (UNISDR 2011). In turn, these extreme weather events are imposing significant costs worldwide, in terms of both human lives and the economy. In agreement with the World Resources Institute, about 100 billion dollars are lost every year due to floods, and while the bulk of these economic costs is concentrated among developed countries, most of the deaths occur in the developing world (IFRC 2014).

Factors influencing floods are related to increased localized and regional precipitation, which further affects river overflows, particularly enhanced by climate change, population and land use changes due to urbanization, deforestation, or agricultural expansion as well as changes in crop types. For instance, growth of the urban fabric toward more hazardous areas increases population vulnerability, especially affecting those social sectors of lower resources that have no other possibilities rather than to settle in floodplains. Flood-induced welfare losses then rebound negatively upon the most vulnerable sectors of society, who often have neither access to public aids nor adequate social safety nets.

Our focal point is to explore the roles of climatic and socioeconomic factors contributing to floods over eastern subtropical Argentina (ESA), a vast region of plains and shallow hills, located to the north of 40°S and east of 65°W in southeastern South America (Fig. 1). The region is highly populated hosting over 60% of Argentina's population, in which agriculture is highly developed, comprising the core region of crop farming in the country. Owing to its geomorphologic, hydrologic, and climatic features, ESA is a territory vulnerable to floods (Latrubesse and Brea 2009). Annual total precipitations widely exceed 1000 mm in most areas (Fig. 2). Major rivers cross the region from north to south, such as the Paraná and the Uruguay, as being part of the southern outflow area of the Río de la Plata basin. The hydrologic cycle shows strong inter-annual and interdecadal variability forced by natural and human-induced climate variations (Antico et al. 2014; Doyle et al. 2012).

It is known that modeled climate projections under global warming, induced by increased greenhouse gas (GHG) concentrations, indicate long-term climate changes and trend changes in the hydrological cycle worldwide, directly impacting on population's welfare (IPCC 2013). The climate change is expected to cause an intensification of the water cycle, which could be manifested by regional steady changes in precipitation, evaporation (evapotranspiration), river discharges, and increased risk of exacerbated extreme phenomena associated with them (Westra et al. 2013). Nonetheless, regional trends and changes in hydrological variables can be upward or downward, depending on the balance between precipitation and evapotranspiration. The latter highly relies on the land cover which, in the case of southeastern South

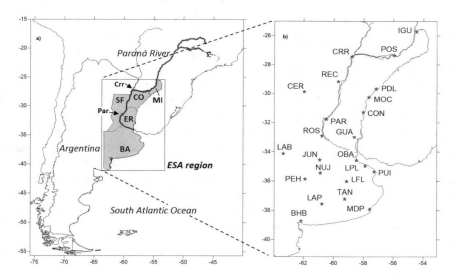

Fig. 1 a Eastern subtropical Argentina (ESA) region, between about 40°–25°S and 64°-54°W in Argentina, southern South America. Provinces of Buenos Aires (BA), Corrientes (CO), Entre Rios (ER), Misiones (MI), and Santa Fe (SF). Markers indicate the Corrientes gauge station (Crr, 27.42°S; 58.85°W) and the Túnel Subfluvial (Tsf, 31.59°S, 60.51°W) along the lower Parana River. **b** Location of meteorological stations within ESA region provided by the National (Argentine) Weather Service. See Table 1 for station names and details

Fig. 2 Annual total precipitation (ATOTP, in mm, intervals every 100 mm) averaged on the period 1961–2010, estimated from monthly gridded precipitation data provided by the Global Precipitation Climatology Centre (GPCP) version 7, on a high 0.5° latitude–longitude resolution grid (Schneider et al. 2011). Inset denotes ESA region

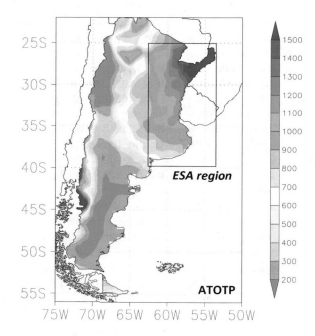

America, including ESA region, has been dramatically altered by the introduction of intensive agriculture (Tucci and Clarke 1998; Krepper et al. 2008).

Therefore, our aim is twofold. On the one hand, we will examine flood risk in ESA region as determined by the maximum flood magnitude (MFM) index devised by the Federal Government. We will assess some changes in the main natural factors related to the hydrological cycle that can cause or increase flood hazard, such as changes in mean and extreme precipitation over ESA, as well as changes in stream-flow and underground water storage during the present climate. We will examine mean precipitation conditions projected for future climate under a business-as-usual scenario of highly increased GHG concentrations. On the other hand, from a sustainable development perspective, we will explore how population has grown in areas with different flood risks, and whether agricultural production has expanded over such areas. To understand people's exposure to flood hazard and land use change is critical to develop integral flood management policies.

Data and Methodology

Spatial flood risk structure and decadal changes of flood risks were explored at the municipality level[1] using MFM index, provided by the Federal Government in the period 1970–2009. This risk index is generated on base of data from the "Desinventar" Project, an inventory system of the effects of disasters, collecting data for each country in Latin America, except Brazil, and a few countries in other continents (India, Iran, Algeria, and Mali). Like EM-DAT, a well-known international disaster database, "Desinventar" relies on official information complemented from other sources, such as newspapers with national coverage. More information about the project is available at http://www.desinventar.org/. The MFM consists of a composite score of flood severity that includes information on number of events and their duration, affected people, and economic losses. Further details on the methodology to obtain the MFM index can be found at the Federal Government Web site "Atlas Indicadores de Desarrollo Territorial de la República Argentina" (http://atl asid.planificacion.gob.ar/). Internal population migrations were also estimated from census data, provided by the National Statistical Office (INDEC, in Spanish), to relate with flood hazards.

Daily precipitation data from meteorological stations provided by the National (Argentine) Weather Service were used (Fig. 1, panel b) to analyze annual total mean precipitation fields and extreme precipitation indices in the period 1961–2015. Extreme precipitation changes were analyzed through linear changes of the total annual precipitation due to strong precipitation days (R75p95pTOT) and due to extreme precipitation days (R95pTOT). The extreme index R75p95pTOT was defined as annual total precipitation when daily rainfall (RR) falls into the

---

[1] Departments or municipalities are the third administrative unit of importance in Argentina below the national and provincial governments.

percentile interval [75th, 95th) of the local daily precipitation distribution of wet days (RR>01.0 mm) in the period 1961–1990, following Donat et al. (2013). Accordingly, R95pTOT was defined as annual total precipitation when RR falls into the percentile interval [95th, 100th). Additionally, to complement annual mean field analysis, monthly gridded precipitation data were used, provided by the Global Precipitation Climatology Centre (GPCP) version 7, on a high 0.5° latitude–longitude resolution grid (Schneider et al. 2011). To assess future climate changes in precipitation and its extremes, under scenarios of anthropogenic climate change due to increasing GHGs, historical and future projections generated from general climate models (GCMs) were analyzed. The simulations corresponded to the Climate Model Intercomparison Project phase 5 (CMIP5. Taylor et al. 2012). We computed multi-model ensemble means for annual total precipitation (ATOTP) and R95pTOT extreme precipitation using 8.5 W m-2 radiative forcing Representative Concentration Pathway (RCP8.5) simulations from the historical (1961–1990) and future climate (2075–2099) experiments. Ensemble means correspond to outputs from four models, selected out of fifteen models which provide daily data (Table A2). The four models, after validation and adjustment for daily frequency distribution bias, achieve to properly represent the historical climate within ESA, as shown by the Argentine report of the Third National Communication to the United Nations Framework Convention on Climate Change (TNC 2014). Simulated daily data are provided on a common 0.5° latitude and longitude resolution grid (retrieved from http://3cn.cima.fcen.uba.ar/).

Long-term changes in monthly river discharges for the Parana at Corrientes and Túnel Subfluvial gauge stations (see Fig. 1), provided by the National Hydrology Agency (retrieved from http://bdhi.hidricosargentina.gov.ar/), were also examined. Monthly gridded data of the terrestrial water storage anomalies (TWSA), as measured remotely by NASA's Gravity Recovery and Climate Experiment (GRACE) mission were examined in the period 2003–2015 (Landerer and Swenson 2012) to complement the hydrological balance in the region. Likewise, information of forest cover and its intra-decadal changes were assessed for Argentina in the period 1990–2015, provided by the UN Statistical Division, Environmental Indicators.

Trends were estimated by means of the linear least square technique (Wilks 2011). The relationship among variables was estimated using Pearson's one-moment correlation (r, and associated explained variance, r^2), and its statistical significance was computed using a Student's t-distribution for a z-transformation of the correlation (Wilks 2011).

Structural Flood Risks

In Argentina, over the past 50 years, 75 major flood events have been reported, affecting around 13 million people and taking more than 500 lives. With the equivalent of USD 22.5bn lost since 1980 (USD 43.5bn after adjusting for the country's GDP growth), floods are the costliest natural catastrophe, representing 58% of economic losses generated by natural catastrophes (Swiss Re 2016).

A crucial aspect is to have an appropriate measure describing the hazard and the vulnerability to determine the extent of flood risks. In this context, we interpret risk as the state of susceptibility to harm from exposure to stresses associated with environmental and social change and from the absence of capacity to adapt (Adger 2006). Thus, flood risk is the combination of flood hazard and people exposed to it. In Argentina, there are no detailed geographical records of flood events and their properties such as extension, intensity, and duration; so, their variability cannot be assessed. Hence, no detailed flood hazard maps are available yet at high geographical resolution.[2] A few engineering studies modeling river dynamics and potential flooding have been carried out, albeit representative of small portions of urban areas (e.g., Saurral et al. 2008). Besides, there are no accessible household surveys to extract information about flooding occurrence or their potential impacts on households' welfare.

Figure 3 presents the mean of the MFM index over the period 1970–2009 for each municipality in Argentina. The map can be interpreted as the structural flooding risk in every municipality. A municipality with an index value of 0 is riskless, at least in terms of flooding, and when the index reaches a maximum of 30, it points to municipalities which are extremely risky in terms of the number of events, duration, material damages, and affected people. Note that ESA is the region where flood risks show the highest figures in Argentina, especially those areas along the Paraná River. For example, those municipalities along the river in the provinces of Santa Fe, Entre Ríos, and Corrientes have a mean MFM index value four times higher than the rest of the municipalities in the same provinces.

Interdecadal variations in the MFM index suggest that the level of hazard has declined between the 1980s and 2000s, the last decade for which data exist, but without reaching previous values of the 1970s (Fig. 4, all municipalities, blue bars). While the all-municipality mean MFM index in ESA in the 1970s and the 2000s was 9.81 and 11.46, respectively, the decadal average index in the 1980s was a staggering 17.39. This could be interpreted as that flood risks have diminished in the region because either local populations have adapted (i.e., by migrating or building resilient infrastructure) or the occurrence of floods has diminished, or a combination of both. This diminution in flood risk is evident for ESA, when excluding the Buenos Aires metropolitan area, known as the Grand Buenos Aires (GBA). The long-term MFM index variation in ESA without GBA (non-GBA) shows an overall negative trend (Fig. 4, red bars and dashed curve). In contrast, the GBA subregion shows a positive trend, since there is a net rise in the MFM index in the 2000s (Fig. 4, GBA, green bars and dashed curve). In the following sections, we will try to further elucidate what factors (climatic features or population exposure) are influencing the observed decadal variation of the MFM index.

---

[2]One way to generate detailed flood mapping for the entire ESA is using remote sensing, such as LANDSAT imagery. Such an endeavor is beyond the scope of this paper, which is left for future research.

Fig. 3 Structural flood risk
determined as the mean
MFM index in the
1970–2009. The municipal
MFM index is produced by
the Argentine Ministry of
Interior. Inset denotes ESA.
Color scale shows the index
magnitude

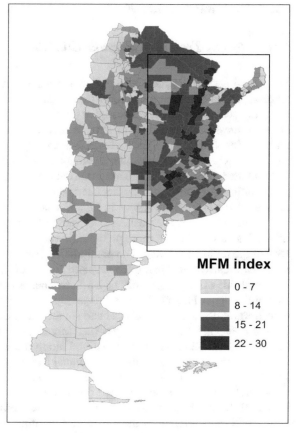

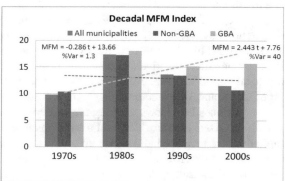

Fig. 4 Decadal mean MFM index estimated from all the municipalities within ESA (all municipalities), estimated without counting the municipalities belonging to GBA (non-GBA), and estimated
only with municipalities within GBA. Corresponding linear trend equations, where *t* is sequential
time. Percentage variance (%Var) explained by the linear trend

Climatic Factors and Floods

Precipitation Trends in Present Climate

The spatial flood risk structure from Fig. 3 is consistent with the spatial mean annual precipitation distribution (compare with Fig. 2), especially over the plains of eastern Argentina, east of the Andes. The region denotes both large annual mean total precipitation values and high values of structural flood risk. Therefore, changes in mean precipitation and strong localized rainfalls can significantly contribute to flooding variations, which in turn may alter the mapping of flood risks.

Estimated from meteorological station data (Fig. 1, panel b), the mean annual total precipitation (ATOTP) and the associated precipitation change due to linear trends in the period 1961–2015 are shown in Fig. 5, panels a and b, respectively. Overall, the precipitation changes are positive over most ESA territory. As a regional average, ESA rains about 150 mm more precipitation over the period 1961–2015, i.e., about 15% of the regional average. The precipitation rises over 500 mm (about 30% of the local historical average) in the northeastern portion of ESA, in the Misiones Province. Note that this northern territory of ESA matches a fraction of the main catchment area of the upper Paraná; thereby, this precipitation increase is influencing the river runoff. Rainfalls also rise in central and southeastern ESA, with increases about 20–25% of the local historical averages (i.e., increases between roughly 150 mm in the central ESA and 250 mm to the east and along the western bank of the Río de la Plata estuary). In the GBA subregion, annual mean total precipitation has increased about 200 mm on average (i.e., about 20% of the historical average).

Several studies have documented that daily extreme rainfalls have substantially increased during the twentieth century in southeastern South America (Cavalcanti et al. 2015, and references therein) and Argentina (Penalba and Robledo 2010), mostly attributable to climate change. To assess the impact of extreme precipitation changes in the region, we examine linear changes of the total annual precipitation due to strong precipitation days (R75p95pTOT; Fig. 5, panels c and d) and extreme precipitation days (R95pTOT; Fig. 5, panels e and f) over the period 1961–2015. The daily precipitation extremes show positive changes over most areas of ESA region, though with sharp local variations (Fig. 5, panels c–f). The spatial distribution of change is like that of the annual total precipitation changes. Strong positive changes are observed over the Misiones Province and along the western shore of the Rio de la Plata estuary, in which GBA is settled. The total annual precipitation due to days with strong precipitation has increased about 90 mm on average over ESA, which represents about 20% of the historical total precipitation due to strong precipitation. The average increase over GBA subregion is about 140 mm, equivalent to over 30% of the historical total precipitation due to strong precipitation, changes comparable to those observed in northeastern ESA.

On a decadal perspective, positive trends in extreme precipitation indices are predominant from the 1960s. Figure 6 shows the decadal mean of annual precipitation totals due to strong precipitation days (R75p95pTOT). Note that discriminating

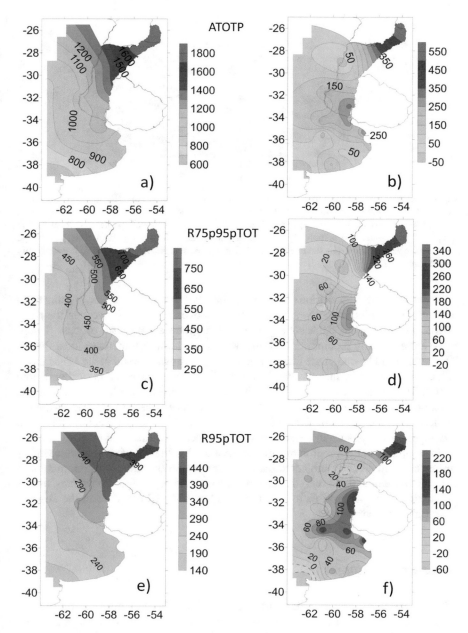

Fig. 5 **a** Annual total precipitation (ATOTP) averaged in the period 1961–2015, estimated from daily station data (in mm, intervals every 50 mm); **b** linear change of ATOTP over the period (N = 55; in mm, intervals every 50 mm); **c** as panel a, but for the annual total precipitation due to strong precipitation days (R75p95pTOT, intervals every 50 mm); **d** as panel b, but for R75p95pTOT (intervals every 20 mm; **e** as panel a, but for the annual total precipitation due to extreme precipitation days (R95pTOT, intervals every 50 mm); and **f** as panel b, but for R95pTOT (intervals every 20 mm). Note that color scales are different for each chart

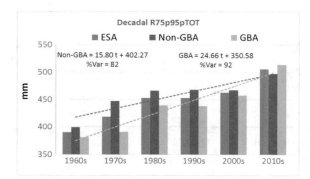

Fig. 6 Decadal mean annual total precipitation due to precipitation days (R75p95pTOT, in mm) averaged across stations in ESA (blue bars), across stations in ESA minus GBA (non-GBA; red bars), and across stations in GBA (green bars). Decade 2010s is proportionally estimated from data in the period 2010–2015. Corresponding linear trend equations, with t the sequential time; percentage variance (%Var) explained by the linear trend. %Var > 45%, significant at two-tail 95%-probability confidence

records for the GBA subregion (green bars), the positive trend is sharper, passing from a mean about 380 mm per year in the 1960s to over 510 mm per year in the early 2010s. Therefore, the GBA subregion has undergone a long-term process toward more extreme daily precipitation events, which would in turn increase local floods. The latter could be one of the factors influencing the upward trend in the MFM index values observed in the last decades for GBA (Fig. 4, green curve). It is worthy to note the GBA recorded one of its worst flash flood events in the last century due to extreme precipitation on April 2 and 3, 2013. La Plata City, in southern GBA, underwent the highest impact in terms of lives lost and damages.[3] Therefore, potential change on the population exposure in GBA subregion is a factor to assess.

In summary, in ESA region in which the national spatial flood risk structure maximizes has been recently subject to increasing precipitation. Thus, the conjunction of both factors, a high flood risk structure and the increasing precipitation values, makes population more susceptible to potential flooding-induced damages in the absence of mitigation management.

River Flow Changes

While extreme precipitation days are crucial to flash flooding episodes on specific localities, in contrast, river flow variations are related to precipitation changes over large areas. We have examined the Paraná River, as representative of rivers in ESA,

---

[3]Precipitation ranged from 150 mm to 390 mm in 24 h within southern GBA, affecting almost 78 thousand homes, causing at least 89 deaths and an economic loss of U$D 666 million in 2013 prices (CEDLAS 2013).

being the second largest in South America. The Paraná crosses ESA from north to south and is a component of the La Plata basin. Its 4,800-km length encompasses a wide basin of roughly 2,600,000 km^2 shared by Brazil, Bolivia, Paraguay, and Argentina. Precipitation over the upper basin (northern ESA, and Paraguayan and Brazilian territories) contributes most to mean annual discharges of the Paraná (Robertson and Mechoso 1998).

Extreme rises of the Parana level are related to the occurrence of El Nino years (positive phase of the El Niño–Southern Oscillation phenomenon), which exert a strong influence on the Parana's inter-annual streamflow variability (Antico et al. 2016). The mean annual streamflow at Corrientes (S) has picked in hydrological year 1982/83 (S=37,819 m^3 s^{-1}), followed by 1997/98 (S=27,267 m^3 s^{-1}) and 2009/10 (S=23,153 m^3 s^{-1}). The last El Nino phenomenon, 2015/16, has contributed with an annual streamflow peak of 27,763 m^3 s^{-1}, all of them further exceeding percentile 95th of the historical distribution (S$_{95p}$=22,976 m^3 s^{-1} in the period 1904/05–2014/15), which shows a historical mean streamflow of 17,225 m^3 s^{-1}. The 1983 flooding was the greatest ever recorded in the past century. Antico et al. (2016) suggested that the 1983 flooding was consequence of the concurrent effect of diverse climate quasiperiodic forcing, such as El Niño and the Pacific Decadal Oscillation, together with the long-term trend related to anthropogenic influence.

The year-to-year variations of precipitation over the middle-upper basin of the Paraná explain about 40–45% of streamflow variations at the Corrientes gauge station. Precipitation variations in the middle-lower basin explain about 30–35% of the inter-annual variation. The latter means that direct precipitation is a weak-to-moderate factor influencing the river streamflow variability at inter-annual scale. Another relevant factor to consider is associated with evapotranspiration, related to groundwater storage. Some experts from the National Agricultural Technology Institute point to limited water storage due to changes in the land cover by agricultural purposes as one of the main factors influencing flooding, as in Bertram and Chiacchiera (2013).

The inter-annual Paraná streamflow variation is highly associated with the terrestrial water storage anomalies (TWSA) from NASA's GRACE mission (Landerer and Swenson 2012) within the wide river basin (Fig. 7). It is quite apparent that the inter-annual Paraná streamflow variation is highly associated with that of the TWSA within the wide river basin (Fig. 7, panel a). Over 80% of the variance of streamflow can be explained by the year-to-year TWSA variability in localities over the upper basin of the Parana. Figure 7, panel b, further shows that there is a positive trend in TWSA over the upper basin of the Parana along the period in which GRACE data are available. Importantly, previous studies have noted the existence of a capacity limitation on terrestrial water storage that is associated with regional flooding (Crowley et al. 2006; Reager et al. 2009). The result points to the high potential risk of Paraná's flooding due to limited water storage capacity in the basin. Furthermore, there is an apparent positive trend in TWSA over the upper basin of the Parana along the period in which GRACE data are available. The latter certainly depends on the rate of deforestation and land changes over the river basin, besides precipitation.

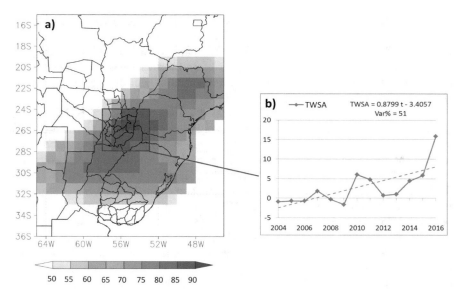

50 55 60 65 70 75 80 85 90

Fig. 7 **a** Percentage variance of the Parana mean annual streamflow at Corrientes gauge station (variance expressed in percentage, color scales), linearly explained by the GRACE's annual TWSA (period 2002/03–2015/16). Variance over 50% is statistically significant at the two-tail 98% level of confidence for TWSA. **b** Annual mean TWSA (in cm) time series averaged in the area 24°–28°S and 58°–53°W (red inset). Corresponding linear trend equations, with t the sequential time; percentage variance (%Var) explained by the linear trend. GRACE land data are available at http://grace.jpl. nasa.gov, supported by the NASA MEaSUREs Program

Interdecadal changes of the mean annual discharges of the Paraná at the Corrientes gauge station, in the lower basin, show an overall tendency toward gradual increase along their records (Fig. 8). Similar behavior is observed at the Túnel Subfluvial gauge station. The change in trend is apparent after the 1960s. Quite like precipitation over ESA, it is evident that the streamflow of the Parana (and other main rivers of the La Plata basin, to which the Paraná belongs to) has strong interdecadal variability forced by known climatic variability such as the PDO (whose negative phase might have induced the streamflow lowering of the 2000s. Antico et al. 2014), though also influenced by global warming (Antico et al. 2016).

Besides climatic natural factors, man-made induced changes such as deforestation (Fig. 9) and global warming might be acting together to contribute to the observed precipitation and runoff trends in ESA that combined with the flood risk structure make the region highly vulnerable. If future potential changes in climate due to global warming persist in the sign of present climate trends, then ESA would become even further vulnerable.

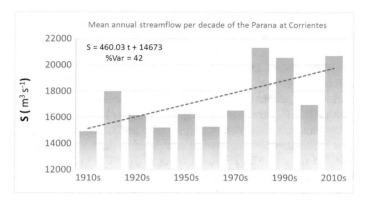

Fig. 8 Decadal mean annual streamflow (S, in m^3 s^{-1}) of the Parana at Corrientes station, and the corresponding linear trend. Decade 2010 s is proportionally estimated from data in the period 2010/11–2015/16. Corresponding linear trend equations, with t the sequential time; percentage variance (%Var) explained by the linear trend. %Var > 27 is significant at a two-tail 95%-probability significance

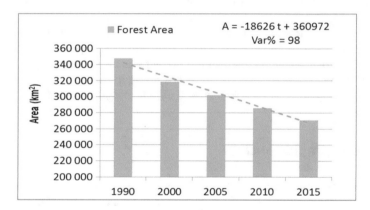

Fig. 9 Forest cover area (in km2) in Argentina from 1990 to 2015 (every five years). Dashed curve indicates linear trend. Corresponding linear equation; A, area; t, sequential time (every 5 years), and percentage variance (Var%) by the linear trend. %Var > 77 is significant at a two-tail 95%-probability significance *Source* UN Statistical Division, Environmental Indicators

Future Projection in Annual Precipitation and Extremes

In climate analysis, GCMs are commonly used to examine and evaluate future climate changes under scenarios of anthropogenic climate change due to increasing GHGs. The CMIP5 has produced a state-of-the-art multi-model dataset designed to improve our knowledge of climate variability and climate change (Taylor et al. 2012). Many researchers have used model output from CMIP5 for analyzing historical and projected climate changes. We have computed four-model ensemble means for ATOTP and R95pTOT using historical and RCP8.5 simulations, which represent the most radical scenario concerning of global.

The historical ensemble means of ATOTP and annual R95pTOT over the baseline 1961–1990 are shown in Fig. 10, panels a and c, respectively. Their percentage changes with respect to baseline, projected by the end of the current century (2075–2099), are shown in panels b and d, respectively. Overall, the historical ensemble means for both ATOTP (panel a) and R95pTOT (panel c) show similar spatial distribution and gradient from west to east and south to north. The percentage changes of ATOTP show moderate change (panel b) compared to the present conditions (panel a). The projected rises in ATOTP are in order of 20–30% in the southern areas of ESA, while in most zones changes are about 10–20% or lower. Figures indicate that the annual precipitation percentage changes show moderate shift compared to the present conditions. The case for extreme precipitation (R95pTOT), in contrast, shows for future projection an overall rise over 30% in most central and southern ESA. Then, it is expected that under a scenario of great GHG emissions, the region would undergo more extreme wet precipitation events than in the present climate.

Note that ESA is the only part of Argentina in which flooding risks due to Parana's runoffs are also projected to increase by the end of the century using GCMs (Saurral et al. 2010; Hirabayashi et al. 2013). Moreover, more frequent and lasting fluvial flooding events in the Paraná and Uruguay basins could be expected using regional climate modeling (Camilloni et al. 2013). Thus, climate projections over ESA by the end of the current century are favorable to increase the vulnerability of flooding, regarding the physical aspect. This exerts a warning to be proactive in developing strategies of adaptation and implementation of new mitigation policies.

Socio-demographic Factors

The climatic factors mentioned above are undoubtedly fundamental in explaining the evolution of recent floods in ESA. However, the vulnerability of the population to such natural hazards is in its essence a multi-causal phenomenon that is also closely related to land use and population changes (Wisner et al. 2003). For instance, unplanned urban sprawl into flood-prone areas, usually observed in less developed countries such as Argentina, results in overall increased vulnerability to a given level of hazard. Moreover, poor people are usually more likely to settle in flood-prone areas because land prices tend to be cheaper (see Rabassa and Zoloa 2016, for evidence on land markets and floods in ESA). Therefore, the welfare costs generated by floods tend to fall disproportionately on these poor sectors of the population because they lack access to reliable formal or informal social safety nets.

Another important factor in explaining overall increase in vulnerability to flooding is the expansion of agricultural production. In ESA, this expansion was achieved at the expenses of native forests and rangelands, especially in the northern lands. As discussed in the previous sections, this process of deforestation due to land conversion could have severely affected evapotranspiration given a certain level of precipitation

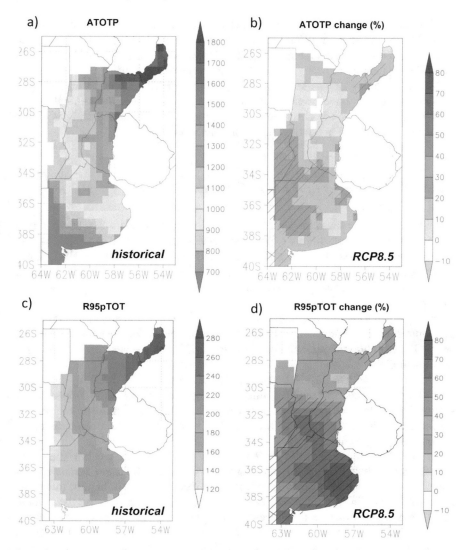

Fig. 10 a Ensemble mean of annual total precipitation (ATOTP, in mm, intervals every 100 mm) from the historical experiment averaged on the baseline 1961–1990; **b** ensemble mean of percentage change in ATOTP (in %, intervals every 10%) from RCP8.5 experiment in the period 2070–2099 with respect to the historical baseline; **c** as panel a, but for annual total precipitation due to extreme precipitation days (R95pTOT, in mm, intervals every 20 mm); **d** as panel b, but the R95pTOT (in %, intervals every 10%). Stripped areas show inter-model spread lower than 10%

making flooding more likely. In agreement with the Argentine Ministry of Environment, Argentina lost 7.6 million hectares of forests between 1990 and 2015, an area the size of Scotland.[4] That means 370,000 hectares of forests that could help absorb water during periods of heavy rain was lost per year (see Fig. 9).

In the remaining of this section, we will explore whether population and agricultural production have been expanding into more risky areas and therefore are contributing factors to the observed vulnerability. We will base our analysis on the MFM index described in Section "Data and Methodology."

Population Changes and Flood Hazards

Recent estimates by the reinsurance company Swiss Re account that about 1 in 3 Argentines lives in areas highly exposed to flooding, amounting to a total of 14.2 m people across the country (Swiss Re 2016), and most of them live in ESA. Thus, understanding changes in population exposure to potential flood hazards is critical to implement an integrated and sustainable approach to floodplain management. As we already discussed, flooding vulnerability is a combination of climatic hazards, which seems to have worsen in recent years, and the economic activity and number of people exposed to those hazards. In this regard, knowledge about the number of people living in floodplains, and how it has evolved in the last decades in Argentina, remains limited.

In agreement with the last National Population Census (2010), in ESA—including the GBA—there exist 213 municipalities which host about 25 million people, representing 62% of the country population. On average, population in those municipalities has increased by 8.28% between 2001 and 2010, with only 15 localities experiencing a decrease in population.

Excluding the municipalities in GBA, those districts in ESA within the upper quartile of the MFM index distribution have approximately 4.5 times as much population as those in the lower quartile of the flood risk distribution. In part, this is the result of the way in which the flood risk index is constructed, because the number of affected people during a flood event tends to be higher in more populated areas. It is for this reason that we cannot infer causality from structural risks to population dynamics. In any case, what seems to be evident from the data, and not surprising, is that population has settled in vulnerable areas along the margins of major rivers.

Perhaps more important, though, is to analyze how population has evolved in the last decade relative to the structural flooding risks. Without disregarding the fact that population dynamics and migration are multi-causal phenomena, and that the deep understanding of them is beyond the scope of this paper, we think it might be useful to explore the spatial correlation between structural flood risks and inter-census population changes.

---

[4]Data available at the UN Statistical Division, Environmental Indicators, http://unstats.un.org/unsd/environment/qindicators.htm.

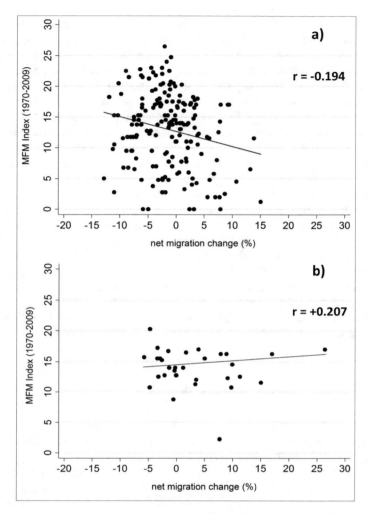

Fig. 11 Dispersion diagram of percentage net migration change between the 1990s and the 2000s and MFM index averaged in the period 1970–2009 (structural risk) for non-GBA municipalities (panel a) and GBA municipalities (panel b). Net migration is the difference between observed population and expected population (as estimated by the INDEC, the Argentine Statistics Office). Negative migration implies net population loss in relation to the expected population (based on birth and death rates). Correlation coefficient, r

We have examined the association existent between the structural risks and the net migration rate[5] for each municipality in ESA, discriminated into non-GBA and GBA municipalities. Figure 11 shows the dispersion diagram between the structural risks,

---

[5]Net migration is the difference between the observed population and the expected population, as estimated the National Statistical Office based on census data.

as measured by the long-term average of the MFM index, and the net migration rate for each municipality in ESA, discriminated into non-GBA (panel a) and GBA (panel b) municipalities. Net migration is the difference between the observed population and the expected population, as calculated by INDEC, the National Statistical Office, based on census data. Though population has increased in most localities, the net migration rate has been negative in many municipalities in non-GBA areas. On average, net migration in non-GBA and GBA municipalities has been -1.07% and $+4.91\%$, respectively. Note that the correlation between net migration and the MFM index for non-GBA municipalities is -0.194 (significantly different from zero at $\alpha=0.01$), implying outmigration from more risky areas. In contrast, the correlation for GBA municipalities is $+0.207$ (significantly different from zero at $\alpha=0.05$), suggesting population gains in more risky localities.

Although it would be quite impulsive to directly attribute a causal relationship between flood risks and migration from correlation analysis, it is not possible to disregard that flooding risks might affect migratory patterns. To our knowledge, no other previous studies have analyzed the issue of migration and floods for ESA, and the exploratory results in this section suggest that it might be an interesting line of research, especially considering climate change and its impacts. The latter might be relevant for the GBA subregion, since such a correlation structure could imply an increase in the number of people exposed to flood hazards, given the projected increased precipitation by the end of the twenty-first century (see Fig. 10, panel d).

Agricultural Expansion and Flood Risks

Besides people's exposure to hazards, one of the main determinants of economic losses due to flooding in ESA is agricultural activity. Every year, the Argentine economy can be expected to lose about 0.15% of its GDP due to flood-related events. While at a first glance, this could appear a minor amount, it adds up to a total annual economic loss of around USD 700 m (Swiss Re 2016).

Agricultural expansion in the region, mostly due to the international demand of soybean and the introduction of herbicide-resistant varieties, has been carefully documented (Viglizzo et al. 2011). The expansion has been at the expenses of native vegetation cover and other crops on already cultivated lands, but principally at expense of the displacement of livestock production into marginal lands. This change in production activities might have an important effect on flood vulnerability (Nosetto et al. 2015).

In ESA, there are 174 municipalities with non-negligible agricultural activities for the 1990s and the 2000s (note that there is no agricultural production in the urbanized GBA subregion). On average, the percentage decadal change in net sown area has increased by 18.05% in those municipalities (66 municipalities had negative change in net sown area). Net sown area is the area which has been sown at least one within any crop season. The massive land conversion into agricultural production could further alter the risk of flooding by reducing soil absorption soon, as pointed by Nosetto et al. (2012). The dispersion diagram between the structural flood risk and the change in net sown area per municipality yields no relevant results (not shown).

Insurance as a Flood Management Policy

It is essential to acknowledge that flood hazard mitigation is a public good, and that the public sector should coordinate actions aimed at reducing risk exposure and promoting adaptation. In Argentina, risk management policies have been historically ineffective in building resilience to natural disasters, and particularly to flood-related shock (Penning-Rowsell 1996; Latrubesse and Brea 2009; Hurlbert and Gupta 2017).

The observed and projected hydrological changes over ESA, depicted in previous sections, suggest that without further human intervention, with new and more effective responses to the flood hazard, human vulnerability will only increase in the future. Increasing resilience to flood hazards should include both structural and non-structural measures. The former includes measures such as stricter regulations and monitoring on land use changes, reforestation and restoration of riparian areas, construction of a resilient transport network, and the development of flood protection barriers, reservoirs and canals for drainage, and pumping systems. Non-structural measures include the design and implementation of meteorological and hydrological early warning systems, the identification and classification of flood-prone areas, and the development of a flood insurance market.

In a context in which most of floodplains are already productively in use or urbanized, as is the case of most of the urban sites along river banks in ESA, perhaps the policy that might be most cost-effective in increasing resilience to flood hazards is insurance. In this regard, the insurance industry in Argentina has very low insurance penetration for the non-life sector, around 2% of GDP in 2015, in accord with Swiss Re (2016). Except for the motor sector and global corporations, where insurance coverage is medium-high, for most other markets flood coverage is either very low or even nonexistent. Standard residential insurance policies generally do not include flood coverage, which would have to be requested by the insured as an ad hoc additional feature. However, very few insurance providers in the country offer this option. This implies that the cost of floods falls entirely on either the affected people or the public sector. For instance, in the abovementioned 2013 event in the cities of La Plata and Buenos Aires, less than 5% of the total damage to households and small businesses was insured (Swiss Re 2016).

Regarding the agricultural sector, around 18 million hectares is insured throughout the country, representing almost half of total cropland (Argentine Farming Risk Office, Federal Government, Web site consulted in 2016). Coverage against flooding is usually only included in multi-peril policies, which represent less than 2% of total insurance coverage, implying extremely low flood insurance penetration in the agricultural sector.

Developing flood insurance markets seems to be a reasonable policy given that floods do not occur frequently in the same area or over vast extensions of ESA, allowing insurance providers to hedge risk effectively across space and time. In this sense, the hindered development of flood-related insurance in the country points to the widespread existence of misconceptions related to the level of exposure to floods. Public opposition to these schemes could be significant, not least because public perception of the hazard threat might be attenuated after several years without major flooding and its consequence (Penning-Rowsell 1996). Besides, once a flood catastrophe occurs, people would assume the government will provide partial financial assistance as usually done in the past.

Conclusions

This research has aimed at exploring the roles of climate and socioeconomic components contributing to floods over ESA, the eastern portion of subtropical Argentina characterized by specific geomorphological, climatic, and hydrological features that make it highly vulnerable to floods. The national structural flood risk map, built upon the MFM index, shows that the region manifests the flood riskiest areas in the country. Within ESA, high flood risk values are observed in those municipalities with the major population settlements. Although decadal mean MFM values have peaked in the 1980s, there is a positive trend of flood risk along the decades over the highly populated GBA, contrarily to what is recorded for the remaining of ESA. This trend of increasing flood risk in GBA is worrisome given the fact that migration rates are higher there, especially in municipalities recording higher structural flood risks.

One major natural factor to consider is flood hazard due to precipitation extreme events. Annual total precipitation has increased roughly 15% on average in the last five decades. A considerable fraction of such an increase is due to more contribution from daily strong precipitation events, which have increased about 30% on average, especially over GBA and northeastern ESA. The upper basin of the Paraná River is subject to increase precipitation which in turn contributes to positive streamflow trend. Deforestation appears to be a high-pressure factor since over 25% of forest land has been since 1990 favoring flooding due to decreased evapotranspiration and limited soil water storage capacity. Future RCP8.5 projections indicate that the hydrological variables will rise by the end of the twenty-first century, particularly exacerbating extreme events of precipitation and streamflow peaks.

Besides these climatic trends and projections over ESA, another critical aspect that makes the region further vulnerable to floods is the lack of continuous and planned flood management policies at the national or subnational levels. For instance, there is no active monitoring system which might be used to implement an early warning system preventing flood damages or flood insurance programs that might help dealing with effective losses. Despite having a national forest law sanctioned in 2007 by the National Parliament to harmonize land use, environmental and socioeconomic issues, the law has not been fully implemented yet due to the lack of allocated funds.

We think that these results, though preliminary in character, can provide policy-makers with a compass to guide future planning to deal with the issue of floods and their impact on sustainable prosperity of the communities. Further research is needed to fully understand the complexity of floods in Argentina.

Acknowledgements This research was funded by the UCA financial support to PEPACG and the "Agencia Nacional de Promoción Científica y Técnica" PICT-2013 0048 Project. Many thanks are given to the Carmelite NGO and Order.

Appendix

Tables 1 and 2.

Table 1 Location of meteorological stations within ESA region provided by the National (Argentine) Weather Service. Stations AER, EZE, LPL, and OBA represent the GBA subregion

Station name	Acronym	Latitude	Longitude
Aeroparque	AER	34.34°S	58.25°W
Bahía Blanca	BHB	38.73°S	62.17°W
Ceres	CER	29.88°S	61.96°W
Concordia	CON	31.30°S	58.02°W
Corrientes	CRR	27.45°S	58.77°W
Dolores	DOL,	36.21°S	57.44°W
Ezeiza	EZE	34.49°S	58.48°W
Gualeguaychú	GUA	33.00°S	58.62°W
Junín	JUN	34.55°S	60.92°W
La Plata	LPL	34.97°S	57.90°W
Laboulage	LAB	34.13°S	63.37°W
Mar del Plata	MDP	37.93°S	57.58°W
Monte Caseros	MOC	30.27°S	57.65°W
Nueve de Julio	NUJ	35.45°S	60.88°W
Observatorio Buenos Aires	OBA	34.58°S	58.32°W
Paraná	PAR	31.78°S	60.49°W
Paso de los Libres	PDL	29.68°S	57.15°W
Pehuajó	PEH	35.87°S	61.90°W
Posadas	POS	27.37°S	55.97°W
Punta Indio	PUI,	35.37°S	57.28°W
Reconquista	REC	29.18°S	59.07°W
Rosario	ROS	32.92°S	60.78°W
Sauce Viejo	SAV	31.42°S	60.49°W
Tandil	TAN	37.23°S	59.25°W
Tres Arroyos	TRA	38.20°S	60.15°W
Iguazú	IGU	25.73°S	54.47°W

Table 2 Model name and institution of the four GCMs from the CMIP5 with daily data used for the historical experiments and projected RCP8.5 experiments in ESA region

Model name	Institution (country)
IPSL-CM5A-MR	Institut Pierre Simon Laplace, France
MPI-ESM-LR	Max Planck Institute for Meteorology, Germany
NorESM1-M	Norwegian Climate Centre, Norway
CCSM4	National Center for Atmospheric Research, USA

References

Adger WN (2006) Vulnerability. Glob Environ Change 16:268–281

Antico A, Schlotthauer G, Torres ME (2014) Analysis of hydroclimatic variability and trends using a novel empirical mode decomposition: application to the Paraná River Basin. J Geophys Res: Atmospheres 119(3):1218–1233

Antico A, Torres ME, Diaz HF (2016) Contributions of different time scales to extreme Paraná floods. Clim Dyn 46(11–12):3785–3792

Bertram N, Chiacchiera S (2013) Ascenso de napas en la Región Pampeana: ¿Consecuencia de los cambios en el uso de la tierra? INTA's technical report N°13. http://inta.gob.ar/documentos/asc enso-de-napas-en-la-region-pampeana-bfconsecuencia-de-los-cambios-en-el-uso-de-la-tierra

Camilloni IA, Saurral RI, Montroull NB (2013) Hydrological projections of fluvial floods in the Uruguay and Paraná basins under different climate change scenarios. Int J Basin Manage 11(4):389–399

Cavalcanti IFA, Carril AF, Penalba OC, Grimm AM, Menéndez CG, Sanchez E, Pántano V et al (2015) Precipitation extremes over La Plata Basin-Review and new results from observations and climate simulations. J Hydrol 523:211–230

CEDLAS (Centro de Estudios Distributivos, Laborales y Sociales, 2013) Medición de Impacto Socioeconómico de las inundaciones en La Plata. http://cedlas.econo.unlp.edu.ar/eng/past-pro jects.php

Crowley JW, Mitrovica JX, Bailey RC, Tamisiea ME, Davis JL (2006) Land water storage within the Congo Basin inferred from GRACE satellite gravity data. Geophys Res Lett 33(19)

Donat MG, Alexander LV, Yang H, Durre I, Vose R, Dunn RJH, Willett KM, Aguilar E, Brunet M, Caesar J, Hewitson B, Jack C, Klein Tank AMG, Kruger AC, Marengo J, Peterson TC, Renom M, Oria Rojas C, Rusticucci M, Salinger J, Elrayah AS, Sekele SS, Srivastava AK, Trewin B, Villarroel C, Vincent LA, Zhai P, Zhang X, Kitching S (2013) Updated analyses of temperature and precipitation extreme indices since the beginning of the twentieth century: The HadEX2 dataset. J Geophys Res Atmos 118:2098–2118. https://doi.org/10.1002/jgrd.50150

Doyle MR. Saurral RL, Barros, V (2012) Trends in the distributions of aggregated monthly precipitation over the Plata Basin. Int J Climatol 32:2149–2162

Hirabayashi Y, Mahendran R, Koirala S, Konoshima L, Yamazaki D, Watanabe S, Kim H, Kanae S (2013) Global flood risk under climate change. Nat Clim Change 3:816–821

Hurlbert M, Gupta J (2017) The adaptive capacity of institutions in Canada, Argentina, and Chile to droughts and floods. Reg Environ Change 17:865–877

IFRC (International Federation of Red Cross and Red Crescent Societies) (2014) World Disasters Report 2014, Geneva

IPCC (2013) Summary for Policymakers. In: Stocker TF, Qin D, Plattner GK, Tignor M, Allen SK, Boschung J, Nauels A, Xia Y, Bex V, Midgley PM (eds) Climate Change 2013: The Physical Science Basis. Contribution of working group I to the fifth assessment report of the intergovern-mental panel on climate change. Cambridge University Press, Cambridge, United Kingdom and New York, NY, USA

Krepper CM, García NO, Jones PD (2008) Low-frequency response of the upper Paraná basin. Int J Climatol 28(3):351–360

Landerer FW, Swenson SC (2012) Accuracy of scaled GRACE errestrial water storage estimates. Water Resour Res 48, W04531, p 11, https://doi.org/10.1029/2011wr011453, 2012

Latrubesse ED, Brea D (2009) Flood in Argentina. Dev Earth Surf Proc 13:333–339

Nosetto MD, Jobbágy EG, Brizuela AB, Jackson RB (2012) The hydrologic consequences of land cover change in central Argentina. Agric Ecosyst Environ 154(1):2–11

Nosetto MD, Paez RA, Ballesteros SI, Jobbágy EG (2015) Higher water-table levels and flooding risk under grain versus livestock production systems in the subhumid plains of the Pampas. Agric Ecosyst Environ 206(1):60–70

Penalba OC, Robledo FA (2010) Spatial and temporal variability of the frequency of extreme daily rainfall regime in the La Plata Basin during the 20th century. Clim Change 98(3):531–550

Penning-Rowsell EC (1996) Flood-hazard response in Argentina. Geogr Rev 86(1):72–90

Rabassa MJ, Zoloa JI (2016) Flooding Risks and Housing Markets: A Spatial Hedonic Analysis for La Plata City. Environ Dev Econ 21(4):464–489

Reager JT, Famiglietti JS (2009) Global terrestrial water storage capacity and flood potential using GRACE. Geophys Res Lett 36(23)

Robertson AW, Mechoso CR (1998) Interannual and decadal cycles in river flows of southeastern South America. J Clim 11(10):2570–2581

Schneider U, Becker A, Finger P, Meyer-Christoffer A, Rudolf B, Ziese M (2011) GPCC Full Data Reanalysis Version 6.0 at 0.5°: Monthly Land-Surface Precipitation from Rain-Gauges built on GTS-based and Historic Data

Saurral R (2010) The hydrologic cycle of the La Plata Basin in the WCRP-CMIP3 multimodel dataset. J Hydrometeorol 11(5):1083–1102

Saurral RI, Barros VR, Lettenmaier DP (2008) Land use impact on the Uruguay River discharge. Geophys Res Lett 35(12)

Swiss Re (2016) Staying afloat. Flood Risk in Argentina. Swiss Reinsurance Company Ltd, Switzerland

Taylor KE, Stouffer RJ, Meehl GA (2012) An Over-view of CMIP5 and the Experiment Design. B Am Meteorol Soc 93(4):485–498. https://doi.org/10.1175/bams-d-11-00094.1

TNC (2014) "Tercera Comunicación Nacional sobre Cambio Climático ante la Convención Marco de Naciones Unidas sobre Cambio Climático". In: Barros V, Vera C, Agosta E, Araneo D, Camilloni I, Carril A, Doyle M, Frumento O, Nuñez M, Ortiz M, Penalba O, Rusticucci M, Saulo C, Solman S (eds) The third Argentine republic communitacion on climate chnange to the UNFCCC. Available at https://ambiente.gob.ar/tercera-comunicacion-nacional/

Tucci CE, Clarke RT (1998) Environmental issues in the la Plata basin. Int J Water Res Dev 14(2):157–173

UNISDR (United Nations International Strategy for Disaster Reduction (2011) Revealing risk, redefining development. UNISDR, Geneva

Viglizzo EF, Federico CF, Lorena VC, Esteban GJ, Hernán P, Jonathan C, Daniel P, María FR (2011) "Ecological and environmental footprint of 50 years of agricultural expansion in Argentina." Glob Change Biol 17(2):959–973

Westra S, Alexander LV, Zwiers FW (2013) Global increasing trends in annual maximum daily precipitation. J Clim 26(11):3904–3918

Wilks DS (2011) Statistical methods in the atmospheric sciences, vol 100. Academic press

Wisner B, Blaikie P, Cannon T, Davis I (2003) At risk: natural hazards, people's vulnerability to disasters, 2nd edn. Routledge, London and New York

Climate Change: Mitigation Policies

Rankings for Carbon Emissions and Economic Growth Decoupling

Mariana Conte Grand

Abstract The main purpose of this chapter is to analyze decoupling between carbon emissions and economic activity for the different countries in the world within the 1990–2012 period. We qualify decoupling cases. Countries are ranked from those that decrease emissions while expanding activity (strong decoupling) to those that augment their greenhouse gases and are in recession (strong negative decoupling). For the cases in which there exists a conflict between growth and emissions (there is improvement in one indicator and worse conditions for the other), the orderings are two, depending if priority is given to the economy or to nature. The findings are that 30% of countries follow green growth paths, 50% weakly decouple their emissions from activity (emissions increase less than GDP), and 20% decouple expansively (their emissions increase more than GDP). There is almost difference between ranking countries giving priority to growth and prioritizing nature. Argentina ranks approximately in the 60th place among around 150 countries in the database and is one of the developing countries that weakly decoupled carbon emissions from economic activity in the period under study.

Keywords Carbon emissions · Decoupling indicators · Degrowth · Green growth · A-growth · Sustainable development

Introduction

As it is well known, the Paris Agreement (PA) main objective is to keep the average increase of global temperature at least below 2 °C with respect to pre-industrial levels by the end of the century in order to avoid massive damages due to climate change. Several research groups analyze the gap between the emissions levels needed to honor that goal and the Parties' climate policies. They conclude that the attainment

M. Conte Grand (✉)
Department of Economics, Universidad del CEMA, Av. Córdoba 374, C1054AAP Ciudad
Autónoma de Buenos Aires, Argentina
e-mail: mcg@ucema.edu.ar

© Springer Nature Switzerland AG 2021
M. E. Belfiori and M. J. Rabassa, (eds.) *The Economics of Climate Change
in Argentina*, The Latin American Studies Book Series,
https://doi.org/10.1007/978-3-030-62252-7_4

61

of the 2° goal requires greenhouse gases (GHG) reductions of 40–70% by 2050, with respect to 2010 (IPCC 2014).

There is a gap between mitigation policies and what is needed to attenuate the consequences of climate change. The gap occurs because that even if the current national commitments under the PA were fully implemented, the target is still far from being reached. In fact, as shown by UNEP (2017), the carbon budget (i.e., carbon allowed emissions by 2100) would be 80% depleted by 2030 (the date of the PA promises). By 2030, there would be a gap to get the world to the 2° trajectory. This happens because greenhouse gases need to be 42 gigatonnes of CO_2 equivalent ($GtCO_2$-e) per year by 2030, and with the PA Nationally Determined Contributions (NDCs), they would be 11–13 $GtCO_2$-e higher.

The only way to grow and fulfill the goal of the Paris agreement is by decoupling economic activities from carbon emissions. According to the Merriam-Webster dictionary, to decouple is "to eliminate the interrelationship of" or "separate." More specifically, OECD (2002) defines decoupling as "breaking the link between environmental bads and economic goods."

The relationship between emissions and GDP evolves differently for each country. This link depends on what and how each of them produces (and consumes). A priori, those economies with a high share of services in value added relative to that of industry or agriculture would be more able to decrease their greenhouse gases to a greater extent, and the opposite would occur for those countries that are major oil producers. However, the profile of each economy is determined by factors that include own initial endowments, technology innovation and changes in consumers' attitude toward the environment. And public policies influence all of them. Decoupling is not automatically attained; it has to be driven by both market and government policy forces (Stavins 2016). And, what is most important in terms of this chapter is that not all types of decoupling are equally desirable.

Different Views About the Relationship Between Economic Growth and the Environment

There are all kinds of difficulties in agreeing on stricter emissions' reduction goals. One of them is that developing countries argue that they are not historically responsible for carbon emissions (i.e., their argument is that concentration of GHG increased substantially since the Industrial Revolution, which began in the developed—and not in developing—nations). Another is that there is no single indicator on which countries can agree on. Per capita emissions and emissions intensity—Emissions/GDP— would be simple metrics to agree on reductions, but that is not possible since they differ substantially between countries: nations with low emissions per capita tend to have high emissions' intensity, but the indicators go the other way around for advanced less populated nations. For example, in 2012 (last data available for total GHG emissions per country at the moment this chapter was written), Paraguay and

Uruguay emitted approximately 8 and 10.1 GHG per capita (tCO_2e/population) while their emissions' intensity was 1089 and 545 tCO_2e/million of GDP measured as PPP (constant 2011 international \$), respectively.[1] Hence, in that scenario, Paraguay would support an agreement based on its emissions per capita while Uruguay would prefer commitments based on emissions per unit of GDP. In addition, there are also conflicts of incentives. Specifically, since GHG mitigation poses local costs but generates global benefits, individual countries have incentives to pursue low levels of effort, expecting that others will take action (this is the well-known "free-riding" problem).

Moreover, there is agreement that higher levels of economic activity favor reductions in extreme poverty, and there is evidence in that respect in the recent world history (Dollar et al. 2013). However, there is no agreement that a higher world GDP is compatible with lower levels of emissions. There are in fact three distinct views referred to the link between growth and nature (Jakob and Edenhofer 2014). One supports "degrowth" as a way to solve environmental pressure on the planet. A second one states that green growth is possible: It is feasible to reduce "environmental bads" and increase "economic goods." A third one favors a-growth. The latter is represented by "growth agnostics": what is valuable is not economic growth, but rather social progress.

The first view is lead by the followers of the Club of Rome that in the 70s recruited scientists from MIT to study the relationship between growth and the environment. They published the result of their research in The Limits to Growth (Meadows et al. 1972). In that work they concluded that if the increase in the world population, the industrialization, pollution, food production and the natural resources exploitation were maintained without any change, the limits to the planet would be attained in the lapse of one hundred years.

Almost simultaneously with Meadows et al (1972), Georgescu-Roegen (1971) used physics to determine that the earth's resources will eventually be exhausted at some point. Georgescu-Roegen argued that all natural resources are irreversibly degraded when put to use in economic activity; consequently, the carrying capacity of earth to sustain human populations is bound to decrease. He based his ideas on the physical concept of entropy. The second law of thermodynamics (or law of entropy) asserts that a natural process runs only in one sense and is not reversible. For example, heat always flows spontaneously from hotter to colder bodies, and never the reverse. Georgescu-Roegen argued not for a second but fourth law by extending the same

---

[1] Total greenhouse gas emissions in kt of CO_2 equivalent are composed of CO_2 totals excluding short-cycle biomass burning (such as agricultural waste burning and Savannah burning) but including other biomass burning (such as forest fires, post-burn decay, peat fires and decay of drained peatlands), all anthropogenic CH4 sources, N_2O sources and F-gases (HFCs, PFCs and SF6). The data is based on estimated emissions, the countries that are Parties to the United Nations Framework Convention on Climate Change (UNFCCC) use to report to the Convention and follow standardized methodologies recommended by the Intergovernmental Panel on Climate Change (IPCC). Note that from the time this chapter was written, only one year (2013) was added for this variable in the World Bank Development Indicators Database. For years from 2014 to 2019 no data for GHG estimators are reported.

idea of energy to resources: when resources are used for human activities, part of them are lost and are impossible to recover. The ideas of that Romanian scientist were closely followed by one of his students (Daly 1973) and were the base of what is nowadays the brand of economics called "ecological economics".

The followers of that rather pessimistic point of view believe that the limits of the planet are getting closer and the solution is to "degrowth" Weiss and Cattaneo (2017) and Cosme et al. (2017) review all publications in this line of literature and affirm that it has been increasing in the last years. According to the latter authors, from the first academic paper that used the term "degrowth" (in 2006) to date, the number of web pages using that word has multiplied by 20. Strictly speaking, the "degrowth" strategy implies not a continuous decrease of economic activities, but rather a transition to a new steady state that considers the limits of the planet. According to ecological economists, the environmental problems can be attributed to an excessively large economy that goes beyond the capacity of nature. They consider there is a problem of scale (Daly 1973). They do not see economic growth as a solution, but rather as a problem. "Degrowth" is a mean to solve the crisis of the planet. As pointed out by Kallis (2011, p. 874), "sustainable degrowth is not equivalent to negative GDP growth in a growth economy. This has its own name: recession, or if prolonged, depression." The "degrowth" solution consists of reaching a new equilibrium, to then allow a growing economy that uses less resources.

The main criticism to the literature of "degrowth" is that the arguments are well described but its feasibility analysis is poor (see in that respect Martínez-Alier et al. 2010; Cosme et al. 2017; Weiss and Cattaneo 2017). Kallis (2011, p. 874) justify this arguing that "degrowth" is an "umbrella keyword.. It has to do with understanding the limits of nature, not to expect technological miracles. Moreover, Jakob y Edenhofer (2014) point out, based on the IPAT identity ($I = P \cdot A \cdot T$ with "I" denoting Impact–emissions—, "P" = Population, "A" = Affluence-per capita GDP—and "T" = Technology -emissions per unit of GDP), reducing emissions 5% annually with a 0.7% increase in population, even if GDP does not change, would require a 5.7%, decrease in emissions intensity, which is not low. They state that it makes little sense to attempt to decrease emissions (I) focusing on decreasing growth (A) when in fact it can be made in a more effective way focusing on other type of policies that emphasize changes in population trends and technology (P and T). Finally, even if it may seem attractive to "degrowth" and live a simpler life, working less hours, it can have negative implications for developing economies, where a minimum material quality of life has not been yet attained (see Martínez-Alier et al. 2010, p. 1743 in that respect).

The second point of view, "green growth", is more optimistic. The term has its origin in 2009, when after the financial crisis the Organization for Economic Co-operation and Development (OECD) published the *Declaration on Green Growth*. In its considerations, it states: "Green growth will be relevant going beyond the current crisis, addressing urgent challenges including the fight against climate change and environmental degradation, enhancement of energy security, and the creation of new engines for economic growth." (OECD 2009, p. 1). Green growth followers believe that it is possible to increase economic activities taking into account the environment.

More precisely, they argue that increases in GDP can have low or even negative costs for nature. With "negative costs" they mean that environmental protection can induce the development of green technologies and businesses that can foster the economy even further. Jacobs (2013) differentiates between "strong" and "standard" green growth. The former is the one for which growth considering nature can encourage the development of growth itself. Hence, followers of green growth are optimistic regarding the feasibility of decoupling between carbon emissions and GDP.

The empirical base for "green growth" is the environmental Kuznets curve, according to which more growth sooner or later implies pollution decreases (Grossman and Krueger 1995). The cause of that inverted U shape between economic activity and emissions is that as countries grow, higher income implies technological development and changes in consumers' tastes, and further both go in the direction of a greener economy. Several articles provide a theoretical framework for the environmental Kuznets curve. For example, Stokey (1998), using a representative consumer model, show that "if increased productive capacity allows both consumption growth and improved environmental quality, then growth may continue without bound." However, there are doubts in the literature on the feasibility of occurrence of a classical Kuznets shape for carbon emissions (Dasgupta et al. 2002).

Finally, other economists support a third way, based on the traditional concept of sustainable development. The idea is that economic growth and environmental sustainability are not goals in themselves, but rather social progress is the ultimate target. Under this line, the economy can grow and do so decreasing emissions, but it can also be that social satisfaction is not increasing. That perspective has been named by the term "a-growth", referring to "agnostic growth." Van den Bergh (2011) introduces that word in the Ecological Economics journal (i.e., one of the most important in the field). "A-growth" does not mean to be against growth, but rather against economic growth that does not consider social and environmental sustainability. An example that is often used to illustrate the point is the case of India that has been able to increase GDP but has maintained a low standard of human development, measured by life expectation, health and education levels (Drèze and Sen 2013).

There are several ways to measure social progress. One alternative is to use social welfare functions (Adler 2012), but it is not easy to operationalize this type of functions since they require comparability among individuals. Another option is to use the index of Genuine Savings (GS), which introduces corrections to take into account environmental resources depletion and environmental damages, as well as the investment in human capital (Hamilton 2000). A criticism for the GS indicator is that losses of natural capital are considered irrelevant if they are substituted by gains in human capital (van den Bergh and Antal 2014). Another index often used to measure progress is the Human Development Index (HDI). The IDH is an indicator that combines life expectation, the number of years in formal education, and GDP per capita. The problem with the HDI is that it does not have any environmental dimension. Arrow et al. (2012) propose another way to measure sustainability in terms of the capacity to provide well being to future generations. They provide a model and empirical estimations of wealth in several countries. They incorporate population growth, technological change, human capital and environmental quality in their measurement of " a comprehensive measure of wealth." Other alternatives to measure progress are happiness or subjective satisfaction indices (Helliwell et al. 2016), or, as proposed by Jakob y Edenhofer (2014) a "dashboard of welfare indicators" as the Sustainable Development Indicators. The problem in this latter case is that, as stated by Fleurbaey and Blanchet (2013), the difficulty in measuring welfare is the multiplicity of indicators. There is no agreement on how social progress can be measured.

Decoupling: Previous Evidence

Beyond any definition of words, there are studies dealing with indicators to actually measure how GDP and carbon emissions decouple. To date, two of them are the most employed. One is the decoupling factor introduced in OECD (2002), defined by the rate of growth of emissions' intensity (emissions/GDP). It states that there is decoupling if emissions' intensity decreases. Unfortunately, it has clear limitations. Decoupling is only associated to a reduction in emissions' intensity, but that scenario can coexist with emissions increasing while the economy is expanding and with emissions decreasing but economic activity falling. The second indicator was introduced by Tapio (2005) and is defined as an emissions-to-economic activity elasticity (rate of emissions' change/rate of GDP change). Depending on the value of this elasticity, there are several types of decoupling scenarios, whose description is the main contribution of Tapio (2005).[2]

Decoupling indicators have been used in several studies to analyze the link between energy, environment and economy. For example, Lu et al. (2007) calculate decoupling in Germany, Taiwan, South Korea and Japan on a yearly base between 1990 and 2003 using the OECD indicator. They find coupling between environmental pressure (transportation CO_2 emissions and energy demand) and GDP except for several years in the first two countries. Freitas and Kaneko (2011), using the same indicator, examine the case of Brazil from 1980 to 2009 and uncover substantial separation between economic activity and CO_2 emissions from energy consumption. Conrad and Cassar (2014) calculate the OECD indicator for several endpoints in the small island of Malta and uncover relative decoupling for greenhouse gases from 1995 to 2011. Gupta (2015) uses that same index to study decoupling for several environmental (not only carbon emissions) endpoints in OECD countries.

Ren and Hu (2012) find different degrees of decoupling for the Chinese nonferrous metals industry in the period 1996–2008 using the Tapio (2005) decoupling index. Zhang and Wang (2013) employ it for decoupling between CO_2 emissions of the whole industry and primary, secondary and tertiary industries in a province of

---

[2] A third measure of decoupling was introduced by Lu et al. (2011) and employed by Wang et al. (2013). Its formula includes, in addition to GDP growth, the emissions' intensity decreasing rate. The three indices can be compared and, in fact, as shown in Conte (2016), Lu et al. (2011) and Tapio (2005) indicators are one a linear transformation of the other.

China (Jiangsu) from 1995 to 2009. A similar analysis is done by Wang and Yang (2015) for carbon emissions in the Beijing–Tianjin–Hebi economic band. Wang et al. (2013) using decoupling indicators for materials use, energy use and SO_2 in China, Russia, Japan and the USA during the 2000–2007 period, conclude that decoupling was stronger in the two OECD nations than in the two BRIC countries because of their different development stages. There are more analysis of this type for different sectors, cities, regions, nations and groups of countries.

In a less academic vein, several think tanks and international agencies evaluate if there is decoupling at the world and at the country level. They assess decoupling without using indicators but by simply looking at the rate of growth of carbon emissions and the rate of growth of GDP. Under this stream, the International Energy Agency, for example, concludes that carbon dioxide global emissions generated by the energy sector have decoupled from the world GDP since those emissions stayed basically stable in the last three years while GDP increased at a 3% rate approximately (IEA 2016). In addition, think tanks as World Resources Institute (WRI 2016) and Carbon Brief (2016) have compared CO_2 emissions and GDP of several countries and conclude that there was green growth (the equivalent of strong decoupling: GDP increases while carbon emissions decrease) for several of them between 2000 and 2013. More precisely, WRI uses CO_2 territorial emissions from the BP Statistical Review of World Energy and GDP (dollars of 2009) from the World Development Indicators for 67 countries. They find that 31% (= 21/67) of the countries in their dataset decreased their emissions between 2000 and 2013 and expanded economically during those years.

For the same period, Carbon Brief (2016) broadened the sample by using production generated CO_2 data from Carbon Dioxide Information Analysis Center (CDIAC) and GDP in each countries' local currency for 181 nations and consumption CO_2 emissions for the same source, which was available for 118 countries.[3] They do so because since it is often argued that developed countries decrease territorial emissions and increase consumption ones, it may happen that decoupling is different when considering consumption and not only production emissions. Peters et al. (2011), for example, show some evidence that rich countries are generally carbon importers (the carbon embodied in the goods they consume is larger than the one of the goods they produce) and the other way around for developing countries. Their conclusions are summarized in Fig. 1. Argentina, for example, is one of the nations that are carbon exporters.

Making that distinction between production and consumption emissions, Carbon Brief (2016) finds that 19% (=35/181) of nations increase GDP while they decrease territorial emissions, and 18% (=21/118) attain green growth when considering consumption emissions. Hence, even if developed countries are carbon importers

---

[3]Less countries keep track of carbon emissions from consumption in part due to the fact that inventories that have to be submitted to the United Nations Framework Convention on Climate Change are production-based.

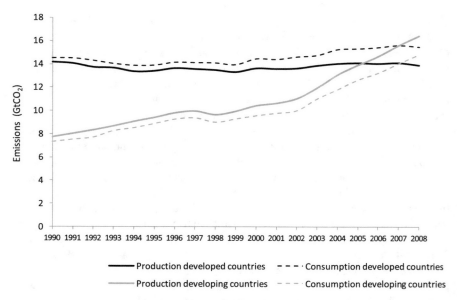

Fig. 1 Consumption and territorial CO_2 emissions in developed and developing countries *Source* Own elaboration based on data in Peters et al. (2011). *Note* Developed countries are those that belong to Annex B of the Kyoto Protocol

(consume more carbon than the one included in what they produce) and developing countries are generally carbon exporters, there is no much difference in the decoupling behavior considering one source of emission or the other.

The question is if green growth is happening and where, but also if a ranking of such decoupling results can be established. The literature on growth and environment centers on the likelihood of a desirable link between economy and the environment; the studies on decoupling indicators stress the types of decoupling they encounter, while the non-academic assessments on decoupling mainly signal those countries that are capable to increase their GDP while decreasing their carbon emissions but do not use indicators. The main innovation of this chapter is to show two decoupling rankings for countries in the world using a decoupling indicator. Instead of discussing which country fits within each type of decoupling pattern, it shows two rankings. Both balance the economy and the environment, but when there is conflict among those two goals, one of them (Ordering I) gives priority to economic growth while the other (Ordering II) prioritizes the environment. The data used for each country come from the World Bank Indicators database, for the 1990–2012 period: more precisely, GDP PPP (constant 2011 international $) and greenhouse gas emissions (kt of CO_2 equivalent). The year 1990 was chosen as the base because it was only in 1992 that the United Nations Framework Convention on Climate Change was signed, and 2012 is the last year for which total greenhouse gases information is almost complete for most countries in the world (if a most recent date is selected, the number of nations that can be included in the analysis decreases significantly).

Decoupling: Concepts and Indicators

The first decoupling indicator introduced in the literature was the one by OECD (2002, p. 19):

$$D_o = 1 - \frac{\frac{E_n}{GDP_n}}{\frac{E_o}{GDP_o}} \tag{1}$$

where E is emissions, GDP is gross domestic product, and the subscripts ($_o$ and $_n$) indicate the beginning and the end of the period respectively. It is straightforward to write D_o as:

$$D_o = -t \tag{2}$$

where t is the growth rate of emissions' intensity:

$$t = \frac{\frac{E_n}{GDP_n} - \frac{E_o}{GDP_o}}{\frac{E_o}{GDP_o}} = \frac{\frac{E_n}{GDP_n}}{\frac{E_o}{GDP_o}} - 1 \tag{3}$$

Then, according to this first indicator, when $D_0 > 0$, there is decoupling because emissions' intensity decreases ($t < 0$). On the other side, when $D_0 \leq 0$, there is no decoupling ($t \geq 0$). Hence, for this indicator, decoupling is synonymous of decreasing emissions' intensity.

Tapio (2005) introduces a decoupling index that refers to the changes in emissions to changes in the economic activity. More precisely:

$$D_\varepsilon = \frac{e}{g} \tag{4}$$

where e is emissions' growth, described as:

$$e = \frac{E_n - E_o}{E_o} = \frac{E_n}{E_o} - 1 \tag{5}$$

and g is the rate of growth of economic activity (usually proxied by the gross domestic product, GDP), characterized as:

$$g = \frac{GDP_n - GDP_o}{GDP_o} \tag{6}$$

According to Tapio (2005, p. 139), there are eight "logical possibilities" (or concepts) depending on the values of D_ε (and e and g). Coupling refers to the situation where D_ε is close to 1 (that is equivalent to saying $e \cong g$). When D_ε departs from 1, there is decoupling. If $D_\varepsilon < 0$ strong decoupling occurs (this means that e and g have opposite signs), if $0 < D_\varepsilon < 1$ decoupling is weak (this implies that e and g have the same sign), and if $D_\varepsilon > 1$, it is just decoupling (and, again e and g have the same sign since $D_\varepsilon > 0$). In the latter case, when both emissions and economy change in the same direction, if they increase this is called "expansive," and when both variables decrease, it is "recessive." Hence, the denomination "expansive" or "recessive" does not come from the value of $D_\varepsilon > 1$, but from the sign of g. The label "negative" is used in all cases when emissions' intensity increases.

To summarize, there are six relevant cases if we discard the very unlikely occurrences in which emissions, GDP and/or emissions' intensity rates of change are zero. Table 1 describes those six scenarios and allows analyzing a couple of interesting features. First, as it is clear from columns 4 and 5, the different indicators have their own values to designate the different kinds of possible coupling/decoupling between emissions and GDP. There are cases for which according to OECD (2002) there is no decoupling (Rows 3, 5 and 6 of Table 1), and there is decoupling for the Tapio (2005) index. Second, neither emissions' intensity decreases nor decoupling (separation between emissions and GDP) are good per se if there are assessed together with the objective of reducing greenhouse gases. It can happen that emissions separate from product while emissions increase (separation of both variables and $e > 0$ in Rows 2, 3, and 6 of Table 1). And, it can perfectly occur that emissions' intensity diminishes at the same time that emissions augment (Row 5 of Table 1, with $t < 0$ and $e > 0$).

Qualifying Decoupling: From Best to Worst Cases

Decoupling cases can be qualified. There is no doubt that the best alternative is strong decoupling, since the economy grows while emissions decrease. Similarly, the worst possible scenario is the one in which GDP decreases and carbon emissions increase. However, if there is conflict and one of the dimensions improves (i.e., or economy increases or emissions decrease), and it is the other way around for the other variable (while GDP increases, emissions increase or while emissions decrease GDP decreases, for example), decoupling cases can still be ranked from best to worse according to two value judgments: prioritize economic growth and put in the second place the environment, or the other way around.

In Ordering I priority is given to the economy, weak decoupling (GDP increases and emissions increase less than GDP) is considered better than recessive decoupling (GDP decreases and emissions decrease more than GDP), but it is the other way around for Ordering II.

More precisely, as a result of Ordering I the order is: strong, weak, expansive negative, recessive, weak negative and strong negative decoupling. For the other

Table 1 Relevant coupling/decoupling cases

e	g	t	$D_o = -t$	$D_\varepsilon = \frac{e}{g}$	Emissions and GDP along time*
−	+	−	$D_o > 0$ Decoupling	$D_\varepsilon < 0$ Strong decoupling	
+	+	−	$D_o > 0$ Decoupling	$0 < D_\varepsilon < 1$ Weak decoupling	
+	+	+	$D_o < 0$ Non Decoupling	$D_\varepsilon > 1$ Expansive negative decoupling	
−	−	−	$D_o > 0$ Decoupling	$D_\varepsilon > 1$ Recessive decoupling	
−	−	+	$D_o < 0$ Non Decoupling	$0 < D_\varepsilon < 1$ Weak negative decoupling	
+	−	+	$D_o < 0$ Non Decoupling	$D_\varepsilon < 0$ Strong negative decoupling	

Source Own elaboration

Note e, g and t are the rates of growth of emissions, GDP and emissions' intensity (emissions/GDP)
*GDP and emissions are considered linear along time only for illustration purposes. The x-axis of the graphs corresponds to time

Table 2 Differences between the two orderings

Order I (priority economy)	Order II (priority environment)
Strong decoupling	Strong decoupling
Weak decoupling	Recessive decoupling
Expansive negative decoupling	Weak negative decoupling
Recessive decoupling	Weak decoupling
Weak negative decoupling	Expansive negative decoupling
Strong negative decoupling	Strong negative decoupling

Source Own elaboration based on Table 1

type of ordering (Ordering II), priority is given to the environment. In that case, the ranking would be: strong, recessive, weak negative, weak, expansive negative and strong negative decoupling. Table 2 shows both rankings: the one that favors economics (Ordering I) and the one that prioritizes the environment (Ordering II).

Types of Decoupling Depending on Some of Countries' Characteristics

With actual data, after calculating the rate of growth of emissions and of GDP (in constant terms) for each country for which there is data available, Ordering I can be made in two steps[4]: (1) Separate countries that grow of those that "degrowth"; and, (2) For the former (g > 0), the smaller the D_ε the greater the decoupling effect and the other way around for countries whose economy decline (g < 0).

As Table 3 shows, 28% of all countries for which there is available data followed the path of green growth (expanding the economy with less greenhouse emissions). In addition, there are 49% nations that decoupled weakly for the whole period: They grew and their emissions increased less than their GDP and only 21% of countries that behaved "wrongfully" (grew emitting more carbon). Table 3 also shows that the link between economy and nature has been more adequate considering Ordering I for the 2000s than in the 90s (82 versus 56% of countries under the strong and weak decoupling scenarios). It is also clear that few nations (3 of 149, namely Tajikistan, Georgia and Ukraine) saw their economy contract between 1990 and 2012. Of the countries that strongly decouple, 62% belong to Europe and Central Asia and those under the expansive negative scenario (the worst for those nations whose GDP increases, under Ordering I) are in sub-Saharan Africa.

Table 4 shows countries organized by decoupling case considering their corresponding World Bank income classification. Even if causality cannot be established, when analyzing the different decoupling degrees between 1990 and 2012 for countries that grow, considering their income level, it becomes clear that nations with

---

[4]It is not enough to use the value of the decoupling indicator. It has to be combined with the rate of growth of GDP (g).

Table 3 Countries in the different decoupling scenarios

		Decoupling case	1990–2000	(%)	2000–2012	(%)	1990–2012	(%)
g > 0	e < 0, t < 0	Strong	32	21	35	22	42	28
	e > 0, t < 0	Weak	54	35	97	60	73	49
	e > 0, t > 0	Expansive negative	49	32	28	17	31	21
g < 0	e < 0, t < 0	Recessive	10	7	0	0	3	2
	e < 0, t > 0	Weak negative	5	3	0	0	0	0
	e > 0, t > 0	Strong negative	3	2	1	1	0	0
		Number of countries	153		161		149	

Source Own elaboration

Note GDP, PPP (constant 2011 international $) and Greenhouse gas emissions (kt of CO_2 equivalent). Both data from World Bank Development Indicators Database. g, e and t denote the rate of growth of GDP, emissions, and emissions intensity (Emissions/GDP), respectively

Table 4 Countries by decoupling behavior and income

Decoupling/income low	Lower-middle	Upper-middle	High		All countries
1990 a 2012					
Strong	1	8	13	18	40
Weak	10	24	20	19	73
Expansive negative	12	6	8	4	30
Recessive	0	3	0	0	3
Weak negative	0	0	0	0	0
Strong negative	0	0	0	0	0
Number of countries	23	41	41	41	146

Source Own elaboration

Note GDP, PPP (constant 2011 international $) and greenhouse gas emissions (kt of CO_2 equivalent). Both data are taken from the World Bank Development Indicators Database. g, e and t denote the rate of growth of GDP, emissions, and emissions intensity (Emissions/GDP), respectively. The number of countries differs from Table 3 because three nations that have GDP and GHG information and so are assigned a decoupling type, are not classified by income by the World Bank: Central African Republic, Congo Democratic Republic and Cote d'Ivoire

higher income levels have been able to strongly decouple carbon territorial emissions from production (44% of high-income nations), but this was not the case of low-income nations (only 4% of them). The contrary occur for less favorable cases as expansive negative decoupling (only 10% of high-income countries and 52% of low income ones). Another way to assess the same effect is that of those nations that are under the best scenario (strong decoupling) 45% have high incomes while only 3% have low income levels.

Around 70% of the countries that strongly decoupled GHG emissions and GDP between 1990 and 2012 reduced the industrial sector share of their economies, and 78% increased the participation of the value added of services. Even if those percentages are high, they also show that GHG-GDP decoupling is feasible in countries with expanding industrial activity. Figure 2 illustrates this point. For example, strong decoupling occurs in the Russia Federation, where the participation of industry

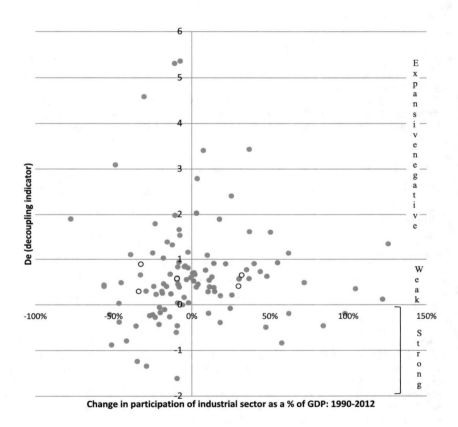

Fig. 2 Decoupling indicator and industrial activity. *Source* Own elaboration. *Note* The countries in this figure are those whose GDP increases from 1990 to 2012, expect for some particular cases as Congo, Dem. Rep., Kyrgyz Republic, Guinea-Bissau, Zimbabwe, which are not included since they have much higher decoupling indicators and so cannot be easily visualized in this figure. The red dot corresponds to Argentina, and the light dots correspond to several of the main Latin American countries: Ecuador and Mexico on the right of the y-axis and Brazil, Chile, Colombia and Uruguay on the left of the y-axis (each one can be identified with the De later reported on Table 5 in this document)

value added (as % of GDP) decreased 35%, but also in Indonesia where that sector increased its contribution to GDP in 11%. Most countries in Latin America (68% of them) weakly decouple emissions from GDP, but not all them have evolved in the same way in terms of their economic structure. As is shown on Fig. 2, in Mexico and Ecuador the share of industry in national value added has increased from 1990 to 2012, while the opposite is true for Argentina, Brazil, Chile, Colombia and Uruguay.

Countries According to Their Decoupling Ranking

Once a decoupling indicator is calculated for each nation, a rank can be assigned according to the two orderings defined above. Table 5 shows each country with its corresponding indicator and hierarchy according to both rankings (from the best to the worst decoupling scenario), for the 1990–2012 period.[5]

Table 5 shows that since there are only three nations (Tajikistan, Georgia and Ukraine) that changed their order because they correspond to cases of recessive decoupling, the correlation among orderings is high (0.94). Hence, even if conceptually, the two orderings are quite different, for these specific data, they rank countries in a similar way.

The Case of Argentina

Argentina has had a significant participation in climate summits over time, and a considerable number of Argentine scientists have been part of the Convention's technical body (the IPCC). There is *expertise* in the country regarding the preparation of inventories and climate studies (see, for example, Barros and Camilioni 2017). Several climate change impacts have been assessed and have to do with the retreat of the glaciers of the Andean mountains, the increase in the frequency of the extreme precipitations (and therefore, of the floods) in the East and Central regions, water stress due to the increase in temperature in the North and West, the increase in sea level affecting the maritime coast and the coast of the Río de la Plata, etc. A possible water crisis is also expected in several provinces of Cuyo and Comahue (MAyDS 2017).

Argentina has been presenting inventories of its greenhouse gases over the years. It has three national communications sent to the UNFCCC (1997 and its revision in 1999, 2008 and 2015) and two biannual emissions reports (2015 and 2017). The

---

[5]Note that there are some countries for which the calculated decoupling indicator is higher than for the rest. This is particularly the case for Congo Dem. Rep., Kyrgyz Republic or Zimbabwe. As noted by Tarabusi and Guarini (2018), this is a characteristic of the decouplingindicator we use: it is not bounded by 1. More precisely, the elasticity is high for these countries as the three ones that belong to our sample and have decoupling indicators above 10 in absolute value. Their particularity is that they have a very low GDP rate of growth for the period studied.

Table 5 Countries with rankings according to the two orderings: 1990–2012

Country name	De	Order I	Order II	Country name	De	Order I	Order II	Country name	De	Order I	Order II	Country name	De	Order I	Order II
Congo, Dem. Rep.	−26.05	1	1	Fiji	−0.06	38	38	Cyprus	0.44	75	78	Guatemala	0.91	112	115
Kyrgyz Republic	−10.77	2	2	Luxembourg	−0.03	39	39	Bahrain	0.45	76	79	El Salvador	0.91	113	116
Cote d'Ivoire	−1.61	3	3	Ireland	−0.03	40	40	Guinea	0.45	77	80	Mauritania	0.93	114	117
Romania	−1.35	4	4	Kazakhstan	−0.02	41	41	Dominican Republic	0.46	78	81	Bahamas, The	0.94	115	118
Russian Federation	−1.24	5	5	Colombia	0.00	42	42	Bhutan	0.49	79	82	Canada	1.03	116	119
Iceland	−1.04	6	6	Vanuatu	0.03	43	46	Tuvalu	0.49	80	83	Togo	1.03	117	120
Bulgaria	−0.88	7	7	Uzbekistan	0.04	44	47	Swaziland	0.52	81	84	Burkina Faso	1.09	118	121
Suriname	−0.84	8	8	Cameroon	0.04	45	48	Nicaragua	0.54	82	85	Malawi	1.11	119	122
Armenia	−0.79	9	9	United States	0.05	46	49	Panama	0.54	83	86	Ghana	1.14	120	123
Czech Republic	−0.65	10	10	Turkmenistan	0.12	47	50	Philippines	0.55	84	87	Barbados	1.15	121	124
Germany	−0.63	11	11	Peru	0.14	48	51	Korea, Rep.	0.56	85	88	United Arab Emirates	1.16	122	125
Denmark	−0.61	12	12	Greece	0.15	49	52	Oman	0.56	86	89	Seychelles	1.32	123	126
Guyana	−0.49	13	13	Malaysia	0.16	50	53	Japan	0.57	87	90	Lao PDR	1.35	124	127
United Kingdom	−0.46	14	14	Puerto Rico	0.20	51	54	Lesotho	0.57	88	91	Dominica	1.39	125	128

(continued)

Table 5 (continued)

Country name	De	Order I	Order II	Country name	De	Order I	Order II	Country name	De	Order I	Order II	Country name	De	Order I	Order II
Papua New Guinea	−0.46	15	15	Bangladesh	0.21	52	55	Chile	0.57	89	92	Djibouti	1.52	126	129
Congo, Rep.	−0.46	16	16	New Zealand	0.23	53	56	Niger	0.58	90	93	St. Vincent and the Grenadines	1.54	127	130
Belarus	−0.43	17	17	Austria	0.24	54	57	Australia	0.59	91	94	Yemen, Rep.	1.61	128	131
Brunei Darussalam	−0.39	18	18	Costa Rica	0.26	55	58	Tunisia	0.60	92	95	Comoros	1.62	129	132
Macedonia, FYR	−0.38	19	19	China	0.29	56	59	Jordan	0.61	93	96	Bolivia	1.67	130	133
Italy	−0.27	20	20	Sri Lanka	0.29	57	60	Vietnam	0.63	94	97	Chad	1.80	131	134
Sweden	−0.27	21	21	Equatorial Guinea	0.29	58	61	Ecuador	0.65	95	98	Gambia, The	1.90	132	135
Paraguay	−0.24	22	22	Uruguay	0.30	59	62	Rwanda	0.66	96	99	Sudan	1.90	133	136
France	−0.24	23	23	Sierra Leone	0.30	60	63	Ethiopia	0.67	97	100	St. Lucia	1.98	134	137
Solomon Islands	−0.22	24	24	Samoa	0.30	61	64	Turkey	0.68	98	101	Mozambique	2.03	135	138
Netherlands	−0.22	25	25	Zambia	0.30	62	65	Thailand	0.70	99	102	Jamaica	2.28	136	139
Albania	−0.21	26	26	Spain	0.33	63	66	Portugal	0.71	100	103	Madagascar	2.40	137	140
Benin	−0.20	27	27	Argentina	0.33	64	67	Antigua and Barbuda	0.71	101	104	Namibia	2.79	138	141
Azerbaijan	−0.19	28	28	Uganda	0.35	65	68	Egypt, Arab Rep.	0.73	102	105	Botswana	3.09	139	142

(continued)

Table 5 (continued)

Country name	De	Order I	Order II	Country name	De	Order I	Order II	Country name	De	Order I	Order II	Country name	De	Order I	Order II
Indonesia	−0.19	29	29	Kenya	0.38	66	69	Israel	0.73	103	106	Senegal	3.41	140	143
Finland	−0.17	30	30	India	0.39	67	70	Algeria	0.76	104	107	Gabon	3.43	141	144
Switzerland	−0.10	31	31	Nepal	0.39	68	71	Tanzania	0.77	105	108	Central African Republic	4.59	142	145
Poland	−0.10	32	32	Honduras	0.40	69	72	Liberia	0.81	106	109	Burundi	5.31	143	146
Angola	−0.10	33	33	Iraq	0.40	70	73	Belize	0.84	107	110	Grenada	5.36	144	147
Norway	−0.08	34	34	Nigeria	0.41	71	74	Pakistan	0.84	108	111	Guinea-Bissau	8.09	145	148
Bermuda	−0.08	35	35	Venezuela, RB	0.41	72	75	Morocco	0.85	109	112	Zimbabwe	28.23	146	149
Myanmar	−0.07	36	36	Mexico	0.41	73	76	Brazil	0.89	110	113	Tajikistan	8.24	147	43
Belgium	−0.07	37	37	Hong Kong SAR, China	0.43	74	77	Mali	0.90	111	114	Georgia	2.86	148	44
												Ukraine	1.91	149	45

Source Own elaboration

Note GDP is PPP (constant 2011 international $) and Total greenhouse gas emissions (kt of CO_2 equivalent). Both data from World Bank Development Indicators Database country is part of those that submitted their "intended nationally determined contribution" (INDC) for the pre-Paris 2015 Summit and is one of the first nations that elaborated a "nationally determined contribution" (NDC) after the Paris Agreement (PA) was signed.

Argentina presented the INDC as a fixed goal (a certain amount of emissions is committed to 2030), but its justification includes a percentage reduction of the projected emissions of a scenario without climate policies, or BAU for *business as usual*).[6] The argument used to support this choice is that this metric is the usual one for developing countries. However, that is not entirely correct. Several of our neighbors did not opt for such type of goals. For example, Chile and Uruguay chose emission intensity targets, while Brazil opted for absolute reductions anchored in the past.

The contribution of Argentina within the PA is that it commits to emit no more than 483 or 369 $MtCO_2$-e depending on receiving or not international financial assistance. This means a reduction of its aggregate emissions by 18% with respect to BAU by 2030 (i.e., 592–483/592), taking emissions without policies whose projection begins in 2005. If Argentina received financial aid to undertake more mitigation measures, total reductions would reach 37% (i.e., 592–369/592).

The text of the NDC states that Argentina's contribution is fair and ambitious considering the data of the last report of the emissions gap (UNEP 2016). The rationale is that Argentina's net emissions in 2014 are 0.7% of global GHG emissions (ie., $369/52,700\ MtCO_2e$) while the country contributes its unconditional target with a 2.8% of the reductions committed by all countries to comply with the Paris Agreement (i.e., 109/3900). For this reason, it is said that their participation in contributions is four times (0.7 * 4 = 2.8) the participation in the emissions. Another statement included in the NDC text is that the unconditional and total reductions are 0.6% and 1.3%, respectively, of the effort necessary to close the gap between the emissions foreseen for 2030 with current policies and those that are needed to reach the goal of the two degrees. However, making an account of the participation in emissions when the unconditional goals of all countries are met, Argentina's participation in 2030 would be 0.87% (483/55,500). This is more than the 0.7% that the country participated in 2014 in the total emissions, but it is the comparison that should be used to justify the fair and ambitious without using the BAU scenario. This would be the correct way if the target, as Argentinean authorities defend, is a fixed target and not one that depends on the BAU scenario.

Despite these justifications, both the unconditional INDC and the NDC of Argentina were considered insufficient by the international community. In particular, a think tank formed by several institutions of international prestige, which is dedicated to monitoring national contributions (see <www.climateactiontracker.org>), stated that, despite the improvements introduced to the goal by the new government, it is "critically insufficient."

---

[6]This has not always been the case. The country designed a contribution that depended on GDP in 1999 (see Barros and Conte Grand 2002).

In fact, Argentina is a country that ranks approximately at the 60th place among almost 150 nations present in the database. In fact, between 1990 and 2012, its GDP increased at an annual average rate of 4%, while emissions increased 2% annually on average (the rates of growth for the overall period are 129% and 43%, respectively). This means that Argentina is decoupling emissions in a weak (and not strong) way. Hence, if the unconditional goal is met, GHG emissions will have increased by 18% between 2005 and 2030 (i.e., 483–409/409, all in GtCO$_2$-e). Only if both types of targets are adopted (unconditional and conditional), would the Climate Action Tracker consider the goal simply insufficient since emissions would fall between 2005 and 2030 by 10% (i.e., 369–409/409).

Argentina has still a long way to go to be able to strongly decouple emissions from GDP. However, it has undertaken several steps that should lead it to a stronger decoupling. In particular, it has established specific plans to reduce emissions in all sectors,[7] Energy policies include: changes in energy efficiency; increase in renewable energy generation; increases in biofuels mixing; changes in large-scale generation as substitution of fossil fuels by natural gas. Transportation emissions' reductions plan encompass: changes in urban mobility through, for example, promotion of public transportation; efficiency improvements in freight transportation; or commercial aviation modernization. Mitigation measures in the agricultural and forestry sectors intend to increase forested areas, limit oilseeds plantation and use biomass for energy generation, among others. Industrial mitigation measures refer to recovery of waste flows, an increase in motors efficiency and the use of more efficient energy sources. In general, policies are weighted in terms of how many emissions they decrease, not in terms of the extent to which they allow decoupling. This happens because there is generally no coordinate work between economic and environment regulatory bodies, and also lack of knowledge of the usefulness of decoupling indicators follow-up.

Conclusions

As shown in this chapter, decoupling greenhouse gases from economic evolution is not favorable per se. It can perfectly happen that emissions and GDP trends separate from each other and emissions increase and/or GDP decreases. Neither augmentations in emissions nor GDP contraction can be an objective to pursue. Similarly, declines of emissions' intensity are not goals by themselves because they can be compatible with increasing emissions and/or GDP contraction.

Hence, "decoupling" as an aim has to be qualified. Orderings can be constructed trying to balance green and growth. Two rankings were constructed here using the decoupling indicator constructed by Tapio (2005). The order depends on, if when there is conflict between economy and nature, priority is given to one or the other.

The results show that around 30% of countries in the world are strongly decoupling their greenhouse gases from their economic activity considering the 1990–2012

---

[7] Refer to https://www.argentina.gob.ar/ambiente/sustentabilidad/planes-sectoriales.

period. This means that in the last several years their GDP increased and emissions decreased. This is the ideal decoupling state. But, there are around 50% more nations that have weakly decoupled (have increased emissions less than economy) and around 20% are in a worse situation. High-income countries tend to have high ranks of decoupling while low income ones usually are on the bottom of the list. In terms of geography, Europe and Central Asia nations are among those that are ranked better and there are many sub-Sahara African countries in the last places.

The two rankings differ for the order they assign: For example, recessive and weak negative decoupling scenarios are considered worse in Ordering I than in II because they imply "degrowth" but emissions that decrease. Since there are only three countries that followed recessive decoupling paths and no countries under the weak negative decoupling scenario, the two orderings do not depart substantially from each other, even if they are conceptually different (one favors economy and the other nature, when there is conflict).

The way the relationship between emissions' and GDP changes evolves differently for each country in the world. This link clearly depends on what and how each of them produces. Desirable decoupling (strong one) is not automatically attained, it has to be driven by both market and government policy forces (Stavins 2016). According to UNEP (2017), there are six areas with alternatives that could be cost-effective to deal with climate change: solar and wind energy, efficient appliances, efficient passenger cars, afforestation and stopping deforestation. Policies within those grounds could imply reductions that would allow fulfilling the Paris Agreement objective. Note that the relevant policies can be climate or non-climate oriented. For example, mercury standards at the US discouraged coal-fired electricity and so reduce GHG emissions. The same happens with motor vehicle standards set because of local air pollution pollutant, but whose efficiency improvement helps climate.

Argentina's climate action has considerably improved in the last few years. Nevertheless, it has only weakly decoupled its emissions from GDP. This means that from 1990 to 2012, country's emissions increase less than GDP. Argentina has compromised an absolute emissions' target at the Paris Agreement. It has also presented specific planned reductions for three sectors (energy, transportation and forestry) and mitigation options in industry and agriculture are under study. Argentina's decoupling behavior may improve in the long-run if the effort underway continues.

Acknowledgements The author thanks participants at the Workshop on Climate Change Economics at Universidad Católica Argentina (April 2018) for their comments, Augustin Shehadi for his help as a research assistant and UCEMA's library personnel for handling all the references' requests.

References

Adler MD (2012) Well-being and fair distribution: beyond cost-benefit analysis. Oxford University Press, p 656

Arrow KJ, Dasgupta P, Goulder LH, Mumford KJ, Oleson K (2012) Sustainability and the measurement of wealth. Environ Dev Econ 17:317–353

Barros V, Camilioni I (2017) Argentina y el Cambio Climático. De la Física a la Política, Eudeba

Barros V, Conte Grand M (2002) Implications of a dynamic target of GHG emissions reduction: the case of Argentina. Environ Dev Econ 7:547–569

Carbon Brief (2016) The 35 countries cutting the link between economic growth and emissions. Policy 5

Conrad E, Cassar L (2014) Decoupling economic growth and environmental degradation: reviewing progress to date in the small island state of Malta. Sustainability 6(10):6729–6750

Conte GM (2016) Carbon emission targets and decoupling indicators. Ecol Ind 67:649–656

Cosme I, Santos R, O'Neill DW (2017) Assessing the degrowth discourse: a review and analysis of academic degrowth policy proposals. J Cleaner Prod 149:321–334

Daly HE (ed) (1973) Towards a steady state economy. Freeman & Co Ltd., W. H, p 332

Dasgupta S, Laplante B, Wang H, Wheeler D (2002) Confronting the environmental Kuznets curve. J Econ Perspect 16:147–168

Dollar D, Kleineberg T, Kraay A (2013) Growth still is good for the poor. World Bank policy research working paper 6568

Dreze J, Sen A (2013) An uncertain glory: India and its contradictions. Princeton University Press, Princeton, NJ, p 434

Fleurbaey M, Blanchet D (2013) Beyond GDP. Oxford University Press, Oxford. xvi + p 306

Freitas de LC, Kaneko S (2011) Decomposing the decoupling of CO_2 emissions and economic growth in Brazil. Ecol Econ 70(8):1459–1469

Georgescu-Roegen N (1971) Entropy law and the economic process. Harvard University Press, Cambridge

Grossman GM, Krueger AB (1995) Economic growth and the environment. Q J Econ 110(2):353–377

Gupta S (2015) Decoupling: a step toward sustainable development with reference to OECD countries. Int J Sustain Dev World Ecol 22(6):510–519

Hamilton K (2000) Genuine saving as a sustainability indicator. In: OECD proceedings: frameworks to measure sustainable development, pp 65–78

Helliwell JF, Huang H, Wang S (2016) The distribution of world happiness. In: World happiness report 2016 update, vol I. Sustainable Development Solutions, New York

IEA (2016). International energy agency. See https://www.iea.org/newsroom/news/2017/march/iea-finds-co2-emissions-flat-for-third-straight-year-even-as-global-economy-grew.html. Accessed 10 Sept 2018

IPCC (2014) Summary for policymakers. In: Climate change 2014: mitigation of climate change. Contribution of working group iii to the fifth assessment report of the intergovernmental panel on climate change [Edenhofer O, Pichs-Madruga R, Sokona Y, Farahani E, Kadner S, Seyboth K, Adler A, Baum I, Brunner S, Eickemeier P, Kriemann B, Savolainen J, Schlömer S, von Stechow C, Zwickel T, Minx JC (eds)]. Cambridge University Press, Cambridge, United Kingdom and New York, NY, USA

Jacobs M (2013) Green growth. In: Falkner R (ed) Handbook of global climate and environmental policy. Wiley Blackwell, Oxford

Jakob M, Edenhofer O (2014) Green growth, degrowth, and the commons. Oxford Rev Econ Policy 30(3):447–468

Kallis G (2011) Defence of degrowth. Ecol Econ 70:873–880

Lu IJ, Lin SJ, Lewis C (2007) Decomposition and decoupling effects of carbon dioxide emission from highway transportation in Taiwan, Germany Japan and South Korea. Energy Policy 35:3226–3235

Lu Z, Wang H, Qiang Y (2011) Decoupling indicators: quantitative relationships between resource use, waste emission and economic growth. Resour Sci 33(1):2–9 (in Chinese)

MAyDS (2017) Cuadernillo Inventario Gases de Efecto Invernadero Argentina. Ministerio de Ambiente y Desarrollo Sustentable. https://inventariogei.ambiente.gob.ar/files/inventario-nacional-gei-argentina.pdf

Martinez-Alier J, Pascual U, Vivien F-D, Zaccai E (2010) Sustainable de-growth: mapping the context, criticisms and future prospects of an emergent paradigm. Ecol Econ 69(9):1741–1747

Meadows DH, Meadows DL, Randers J (1972) The limits to growth. Universe Books, New York, NY

OECD (2002) Indicators to measure decoupling of environmental pressure from economic growth. Sustainable Development. SG/SD (2002) 1/Final. Organization for economic co-operation and development

OECD (2009) Declaration on green growth adopted at the meeting of the council at ministerial level on 25, C/MIN(2009)5/ADD1/FINAL

Peters GP, Minx JC, Weber CL, Edenhofer O (2011) Growth in emission transfers via international trade from 1990 to 2008. Proc Nat Acad Sci 108:8903–8908

Ren S, Hu Z (2012) Effects of decoupling of carbon dioxide emission by Chinese nonferrous metals industry. Energy Policy 43:407–414

Stavins RN (2016) The ever-evolving interrelationship between GDP and carbon dioxide. Environ Forum 33(4):17

Stokey NL (1998) Are there limits to growth? 39(1):1–31

Tapio P (2005) Towards a theory of decoupling: degrees of decoupling in the EU and the case of road traffic in Finland between 1970 and 2001. Transp Policy 12(2):137–151

Tarabusi EC, Guarini G (2018) An axiomatic approach to decoupling indicators for green growth. Ecol Ind 84:515–524

UNEP (2011) Decoupling natural resource use and environmental impacts from economic growth. United Nations Environment Program. A report of the working group on decoupling to the international resource panel

UNEP (2016) Emissions gap report 2016. United Nations Environmental Program

UNEP (2017) Emissions gap report 2017. United Nations Environmental Program

Van den Bergh JC (2011) Environment versus growth: A criticism of "degrowth" and a plea for "a-growth." Ecol Econ 70(5):881–890

Van den Bergh JC, Antal M (2014) Evaluating alternatives to GDP as measures of social welfare/progress, WWW for Europe working papers series 56

Wang H, Hashimoto S, Yue Q, Moriguchi Y, Lu Z (2013) Decoupling analysis of four selected countries: China, Russia, Japan, and the United States during 2000–2007. J Ind Ecol 17(4):618–629

Wang Z, Yang L (2015) Delinking indicators on regional industry development and carbon emissions: Beijing-Tianjin-Hebei economic band case. Ecol Ind 48:41–48

Weiss M, Cattaneo C (2017) Degrowth-Taking stock and reviewing an emerging academic paradigm. Ecol Econ 137:220–230

WRI (2016) The roads to decoupling: 21 countries Are reducing carbon emissions while growing GDP. World Resources Institute

Zhang M, Wang W (2013) Decouple indicators on the CO_2 emission-economic growth linkage: the Jiangsu Province case. Ecol Ind 32:239–244

"Multiple Dividends with Climate Change Policies: Evidence from an Argentinean CGE Model"

María Priscila Ramos and Omar Osvaldo Chisari

Abstract Given the international commitments concerning Climate Change, we evaluate the costs and the potential multiple dividends of applying a carbon tax and/or an environmentally oriented trade policy (EGA) as part of an Environmental Tax Reform (ETR) in Argentina. Reviewing the literature concerning the conditions under which multiple dividends of an ETR can emerge, and also comparing the dividends and costs of implementing alternative ETRs in Argentina using computable general equilibrium model simulations, we conclude that given the structural socio-economic characteristics of developing countries (*i.e.* persistent unemployment, uneven income distribution, recurrent external and fiscal imbalances, high capital volatility, among others), the implementation of an ETR could be more expensive than for countries without these constraints. For instance, an ETR that leads to an increase in the unemployment rate would easily become impracticable. Results highlight that multiple dividends could emerge when a carbon tax helps to reduce distortionary taxes on labour. Unemployment due to fixed real wages and low capital mobility across sectors and countries allow for this multiple-dividend result. Moreover, when this first best choice of ETR is not applicable, an EGA could also lead to multiple dividends under the same labour market conditions, but with greater capital mobility across sectors and the possibility to import a greener production technology. However, the latter could exert a high pressure on the external balance when implementing a foreign technology. Designing an ETR for developing countries requires this cost-dividend analysis since results seem to be highly sensitive to factors market conditions and the available technology.

M. P. Ramos (✉)
Departamento de Economía, Facultad de Ciencias Económicas, Universidad de Buenos Aires, Buenos Aires, Argentina
e-mail: mpramos@economicas.uba.ar

M. P. Ramos · O. O. Chisari
CONICET-Universidad de Buenos Aires, Instituto Interdisciplinario de Economía Política (IIEP-Baires), Buenos Aires, Argentina
e-mail: ochisari@gmail.com

M. P. Ramos
Centre d'Études Prospectives et d'Information Internationale (CEPII), Paris, France

© Springer Nature Switzerland AG 2021
M. E. Belfiori and M. J. Rabassa, (eds.) *The Economics of Climate Change in Argentina*, The Latin American Studies Book Series,
https://doi.org/10.1007/978-3-030-62252-7_5

Keywords Climate change · Carbon tax · Environmentally oriented trade policies · Dividends · CGE model · Argentina

Introduction

International coordination is important for the reduction of greenhouse gas (GHG) emissions. Left alone, most countries would not have incentives to fight climate change (CC) and would rather allocate resources to adaptation taxes (Chisari et al. 2016). That is rational at an individual level, but inefficient from a social (world) perspective.

Thus, international agreements and their effective enforcement seem to be unavoidable. More than 200 multilateral environmental agreements ratified by, at least, more than two countries have been signed in the last 20 years. However, only a few of them have been oriented to seek coordinated actions facing the CC problem. Among them, we can mention: the Montréal Protocol (1985); the Kyoto Protocol (1997); and the recent Paris Agreement (2015). For instance, the Paris Agreement seeks to achieve a global and common commitment by all countries but differentiated based on both their responsibility in the context of global emissions and their financial possibilities and development constraints. Nevertheless, none of these agreements explicitly states the obligatory nature of the means to achieve the emission reduction targets or global temperature targets (maximum increase of 2 °C according to the Paris Agreement). For that reason, each country should design its own environmental instrument which best suits the state of its economy.

Argentina, like other Latin American countries, is a small country in terms of its GHG emissions—less than 1% of the global carbon dioxide (CO_2) emissions are due to this economy—particularly compared to the USA, China and the European Union, which account for more than half of the global CO_2 emissions (CAIT Climate Data Explorer 2015). Even though the comparative responsibility of Argentina for the global GHG emissions is minor, efforts are not costless for the economy (*e.g.* lower consumption, changes in resource allocation, need for technological reconversion in some sectors and change in consumption behaviours and preferences). Moreover, the costs of reducing GHG emissions are greater for the economies under macroeconomic stress, as is the case of Latin American countries. Thus, it might not be wise to constrain the discussion of CC policies to the normative dimension that designs optimal economic policies under ideal conditions of full employment, well-behaved markets and abundance of capital. Instead, the recommended CC policies for developing countries must be reconsidered through the lens of their structural problems.

Consequently, setting the current challenge for researchers and policymakers, the choice of a CC policy instrument (*i.e.* the carbon tax, cap-and-trade measures and other second best tax packages) has to take into account the functioning of the factors market (unemployment, wage rigidities, capital flight), the external restrictions (low diversification of exports, recurrent balance of payment crises), structural social issues (poverty and inequality) and government and institutional characteristics (fiscal deficits, tax evasion, chronic corruption, deficient public goods provision, lower stringency of law and weak policy enforcement) of developing countries, such as Argentina.

In this sense, CC policies in Argentina should be tackled as part of a wider sustainable development programme, where the CC mitigation should not deteriorate (but improve, if possible) other structural constraints pervasive in developing countries, as mentioned above. Thus, instead of limiting the discussion to the environmental benefits of environmental or trade policies, we will enlarge the analysis considering whether other benefits or losses arise in the fight against GHG emissions. It is true that the literature on environmental policies has been exploring the presence of other dividends for the economy, with and without a reform in the national tax system, mainly a double dividend. In general terms, the double dividend literature studies if, additionally to environmental gains, efficiency gains of the tax system can be attained by replacing distortionary taxes with carbon taxes. We will explore the existence of other dividends that are also critical, such as the impact of emission taxes and trade policies on the trade balance, unemployment, and income distribution.

Notwithstanding the foregoing, at the end of the day, all dividends could be reduced to only one index: welfare. Welfare is directly impacted by climate change (*e.g.* discomfort, health problems and impoverished landscapes), and indirectly, for example, through the reduction of productivity and the loss of endowments (*e.g.* arable land). However, it is difficult to synthesize everything in only one dimension due, among other reasons, to differential intergenerational and intra-generational effects of shocks and policies. Beyond that, macroeconomic distress can influence the evaluation of welfare, particularly under conditions of unemployment of resources. That is why it is useful to consider several dimensions for potential gains and losses.

This chapter aims at analysing different CC policies under the factors market behaviour that characterizes the Argentinean economy. After analysing the literature on multiple dividends of environmental policies (Sect. "Literature on Multiple Dividends of the Environmental Policy") and describing the environmental concerns of Argentina in the current context (Sect. "Argentina and the CC Commitments"), we present a Computable General Equilibrium (CGE) model for this country, assuming unemployment, capital restriction and rigidities in both factor markets, as a tool for simulating counterfactual CC policy scenarios (Sect. "Appropriate Tool for Measuring Dividends of CC Policies: A CGE Model for Argentina"). For policy comparison, we have chosen the results from the implementation of a carbon tax in Argentina from Chisari and Miller (2015) and the active participation of Argentina in the plurilateral liberalization of environmental goods and services (EGS) from Ramos et al. (2017). Even though, from a CC perspective, the carbon tax is the first policy option and the trade policy is a second-best, we evaluate not only the reduction in carbon emissions, but also other positive side effects on development variables (employment, income distribution, trade and pressure of the external balance, GDP). Final remarks concerning the Argentinean CC policy could also contribute to the discussion and the design of CC policies for other developing countries in Latin America.

Literature on Multiple Dividends of the Environmental Policy

Both theoretical and empirical literature on environmental policy agrees that the first dividend of, for instance, a carbon tax is a cleaner environment, which, applied to our case of study, implies lower GHG emissions (de Mooij 2000); however, the evidence is not conclusive concerning other non-environmental dividends.

According to the meta analysis of Freire-González (2018), who revised 40 empirical papers using CGE models (1993–2016) looking for economic dividends of an **environmental tax reform (ETR)**, the emergence of a second dividend (improvement in economic efficiency) and a third dividend (greater employment) depends on the current tax structure (optimal or not) and on the instruments through which the carbon tax revenue is recycled (labour tax, capital tax, income tax,[1] food taxes/tariffs, lump-sum transfers), agents' preferences, factors mobility (sector-specific factors in polluting sectors, mobility across sectors, regions and international mobility), factors substitution (energy consumption vs. capital of new technologies), institutional enforcement,[2] among others. He also points out that the double and triple dividends appear mainly in European countries, and even though Africa, Oceania and South America also confirm this expected result, it is not sufficiently supported statistically because of the lack of evidence from these continents.

Moreover, Freire-González (2018) notices that the pollution tax must be dynamic given that this instrument motivates the reduction in pollution and, thus, the tax base; consequently, the switch to other pollutants as pollution tax base should be necessary in order to keep environmental and economic dividends over time. The main conclusion of this paper, which defines the forthcoming research on CC policies, is the need for the design of an ETR taking into account the particularities of each economic system in order to yield not only the environmental dividend, but also other multiple economic dividends simultaneously.

In the next subsections, we summarize the discussion about the traditional double dividend that concerns the recycling of a pollution tax through other distortive taxes in order to reduce the deadweight losses and to increase the GDP. Then, we introduce the possibility that an ETR leads to other dividends linked to the labour market (unemployment) and socio-economic conditions (poverty and income distribution). Finally, we discuss the impact of CC policies on the external balance, particularly when trade, FDI and international cooperation for technological innovation transfers are allowed. The chosen papers for this literature review meet the applications of developing countries and the main behavioural assumptions for the modelling of developing countries.

---

[1] Allan et al. (2014) finds that only income taxes with forward-looking behaviour of agents secure the double dividend (lower carbon emissions and greater GDP) in Scotland.

[2] Castiglione et al. (2018) finds econometric evidence of institutional issues (rules of law, policy enforcement) as key determinants of an ETR in the European Union.

The Double Dividend: Lower Emissions and Greater GDP

Together with trade policies, tax policies are the preferred tools that provide incentives to limit GHG emissions. In fact, it is highly probable that taxes will play a fundamental role for CC policies in the future since they are more easily administered by governments from developing countries, such as Argentina (Aldy et al. 2010). More sophisticated instruments, like cap-and-trade mechanisms, could be more demanding in terms of the supply of institutional services that those economies can provide. There is agreement then with Tol (2008), who argues that taxes on emissions are the lowest cost instrument.

When only taxes are taken into account, then we have to tackle the question of how to define double or multiple dividends. According to Schöb (2003), the "weak" form of the *double dividend hypothesis* states that a revenue-neutral green tax reform is able to cut distortionary taxes, thus lowering the efficiency cost of the green tax reform. The "strong" form of the double dividend asserts that a green tax reform not only improves the environment, but also increases non-environmental welfare.

Giménez and Rodríguez (2010) argue that this public finance approach, which differentiates weak from strong double dividends of an ETR on the basis of efficiency gains/losses of the tax system, is not an appropriate measure to evaluate the double dividend under a general equilibrium (GE) approach. Since the GE models allow for interactions between taxes, it is necessary to isolate each tax change in welfare, *i.e.* first, the change in welfare due to the environmental tax compared to the benchmark (without environmental taxes), and then, the change in welfare due to the reduction of other distortionary taxes after introducing the pollution tax.[3]

In the particular case of Latin American countries, there is a shortage of empirical studies about the double dividend. Chisari and Miller (2015) find a rise in the double dividend in Argentina, Chile, El Salvador, Jamaica and Peru, but not in Brazil, when reducing labour taxes to compensate the additional revenue produced by the carbon tax. Grottera et al. (2017) analyse the case of the carbon tax in Brazil, without and with recycling tax revenue through lower labour tax or greater lump-sum transfers to poor households. In this case, the double dividend emerges only with lower labour tax compensation. No change in technology is allowed in the previous results. But allowing for technological innovation in the energy sector exogenously (following the historical trend) and endogenously (through a greater substitution between capital and energy when prices change due to carbon tax), Rivera et al. (2016) find an unambiguous double dividend in the case of Mexico when achieving its CO_2 emission targets by 2050. This result is also reached through an ETR that allows for the compensation of the revenue from distortionary taxes with the introduction of a carbon tax.

---

[3]Giménez and Rodríguez (2010) compare double dividend measures for the ETR in the USA.

The Third Dividend: Less Unemployment and Less Inequality

The previous double dividend of an ETR might become a modest objective for developing economies. Therefore, other dimensions of socio-economic dividends have to be taken into account. For instance, carbon taxation results in the redistribution of income and impacts on welfare too. Those changes are important and could trigger political opposition that could block its use. In that sense, the determination of wages (constant in real vs. nominal terms) and capital mobility are key elements in assessing the quantitative impact of carbon taxes (Chisari and Miller 2015). In this sense, we have to calculate the best ETR under unemployment conditions. And for this particular issue, it should be recalled that second best Ramsey taxes must be corrected reducing those applied on goods which are labour intensive (Marchand, Pestieau and Wibaut, 1989; Koskela and Schöb 2001; Böhringer et al. 2005). In this regard, Carraro et al. (1996) show that the employment dividend could be generally reached when the carbon tax replaces labour (social security) taxation. Moreover, Chisari and Miller (2015) find that, under full employment, the cost of lower emissions is significantly reduced with respect to the unemployment case. These results raise the point that economic dividends are sensitive to the conditions specified for the labour market, in particular, and the factors markets, in general.

Bovenberg and Goulder (2002) additionally find that this third dividend is particularly achieved when the sector levied by the carbon tax is not labour intensive; otherwise, employees are the ones who assume directly (or indirectly through good prices and unemployment) the cost of the carbon tax (Layard et al. 1991; Koskela and Schöb 1999). Moreover, under unemployment in a small open economy, the employment dividend could appear if labour is highly substitutable with other resources, if the participation of labour in value added is large, and if the relative taxation between labour and other factors is initially high (Bovenberg and Van der Ploeg 1994).

Ciaschini et al. (2012) find that, even when the employment dividend could be reached at the national level, regional disparities could intensify depending on tax recycling. They compared the impact of two tax recycling scenarios (national income tax and regional taxes) of an ETR between the north and the south of Italy, and they find that a lower unemployment rate is achieved only in the north but, unfortunately, not in the south, where this scourge is more severe. Kolsuz and Yeldan (2017) also observe that promoting green employment through carbon tax revenues could lead to a triple-win situation for GHG emission abatement, employment and economic growth in Turkey.

The introduction of structural characteristics in the dynamic CGE model makes it possible to infer dividend relations applicable to other developing countries. For example, the presence of dual economies (formal and informal) and urban–rural migration could modify the expected results. Markandya et al. (2013) empirically demonstrate for Spain that, with a shadow economy (informal labour), a carbon tax allows for the reduction of inefficiencies indirectly taxing this informal labour through greater prices. Under this context, when a carbon tax compensates lower (formal) labour taxes, informal labour becomes formal, GDP increases and the unemployment rate also falls, thus leading to a triple dividend. Kuralbayeva (2015) proposes to consider a triple dividend of an ETR associated not only with unemployment, but also to rural–urban migration. In the case of Mexico, but also applied to other developing countries, Kuralbayeva (2018) finds that, under these particular labour market conditions (unemployment, informality and regional mobility), a carbon tax (urban sectors) could lead to a reduction in the unemployment rate through the labour migration from urban to rural areas that depresses labour incomes in the latter.

An ETR could also help to fight inequality or poverty depending on the change in the purchasing power of low-income households. When the initial tax system is non-optimal from a non-environmental point of view, a carbon tax could have a regressive impact on income distribution if the pattern of consumption of poor households is intensive in carbon-intensive goods; however, this inequality effect could be compensated by a tax recycling option through other taxes (Klenert et al. 2016). This effect could be produced when the tax reduction focuses particularly on improving food access. Thus, the possibility of reaching a triple dividend—decreasing emissions, increasing GDP and decreasing poverty—emerges when the carbon tax makes it possible to reduce food prices (Van Heerden et al. 2006).

As mentioned above, Grottera et al. (2017) also focus on the triple dividend in Brazil, looking for a tax recycling scheme through an ETR that leads to lower levels of inequality. Lump-sum transfers to poor households do not result in a triple dividend, but lower labour taxes, *i.e.* moderate carbon taxes (25R/CO_2$t), and lower labour taxes not only lead to lower carbon emissions, but also lower inequality and unemployment.

The Fourth Dividend: Improving the External Balance

A strong double/third dividend could be blocked when it creates trade balance disequilibria. Thus, it is also important to look at a fourth dividend of an ETR, which concerns the improvement of the external balance, since it is also a weakness of developing economies, and Argentina is not an exception.

Given the multiple dividends we are looking for, a single policy instrument would not be enough and, instead, it would be necessary to continue exploring different combinations of environmentally oriented policy instruments for a full ETR.

Chisari and Miller (2015) find that the costs of reducing emissions in small economies are not negligible due to two main reasons: the state of the labour market (analysed above) and the impact on exports. As mentioned above, the implementation of carbon taxes leads to magnified costs when wages are not determined in the market and when capital is freely mobile with respect to the rest of the world. But, additionally, when real wages are inflexible downward, higher costs due to taxes also increase nominal wages. Consequently, domestic goods become less competitive and exports are thus reduced, decreasing the activity level of firms and the total employment rate. Capital flight due to differentials in the rate of return with respect to the rest of the world magnifies the reduction of the domestic activity. They find that the presence of an alternative cleaner technology, which competes with the incumbent technology, helps to reduce the costs in a significant way since it is not too demanding on foreign resources. Therefore, the less stress a low-emission technology exerts on the external accounts of an economy, the likelier it will be voluntarily adopted.

Argentina has a similar production and trade pattern to that of Australia, so the conclusions reached by Fraser and Waschik (2013) concerning economic dividends dependent on carbon tax bases in Australia could be useful to design an ETR in similar economies, such as Argentina. They confirm the conclusion by Bento and Jacobsen (2007): the higher the participation of specific factors in carbon-intensive sectors, the greater the double dividend (lower GHG emissions and greater welfare) when a carbon tax is applied. This assumption is realistic in energy sectors where sunk-specific capital prevails. Environmental and welfare dividends become stronger when the tax base is on production and not on the end-use because of the path-through difference in the price change. Trade balance and carbon leakage are also highlighted as other spillover effects of the carbon tax applied to different tax bases. The production tax base reduces exports of carbon-intensive sectors but incentivises the consumption of carbon-intensive imported goods, leading to potential leakage increases. In contrast, the consumption tax base of a carbon tax reduces the demand of carbon-intensive goods (domestic and imported), leading to a potential reduction in production because of a lower domestic demand and an increase in exports, which are generally lower than domestic sales because of the Armington assumption on preferences. To avoid the overestimation of the carbon leakage results, it is necessary to allow for technological innovation and its positive spillover effects on the CC modelling. This conclusion is also theoretically and empirically (Switzerland) supported by Karydas and Zhang (2017), who demonstrate that, in the long run, a carbon tax could lead to the development of the R&D sector, resulting in technological change and, thus, an endogenous green growth. Additionally, Fraser and Waschik (2013) consider an ETR that also includes a tariff reform when carbon leakage really arises as a consequence of the application of a carbon tax. In that case, it would be necessary to address this issue through border trade policies, such as greater tariffs on carbon-intensive goods.

In line with the previous idea, some joint trade and environmental negotiations look for environmental and external dividends. However, as evidenced, the addition of growth and socio-economic dividends (lower unemployment and poverty) seems to be very challenging goals for developing countries.

Other types of ETR scenarios linked to more pragmatic trade policies appear in the context of Environmental Goods and Services Negotiations, either in multilateral (the Doha Round) or in plurilateral (e.g. APEC) negotiations. For instance, these ETRs seek to increase relative prices of energy-intensive and polluting goods, through the increase in their tariffs (Fontagné and Fouré 2016a, b), the reduction of the export rebates related to these products (Fan et al. 2015) or else the reduction of tariffs on environmentally preferable products (Gozlan and Ramos 2008) or on low energy-intensive goods (Ramos 2014; Ramos et al. 2017). These trade policies appear as second-best alternatives against the CC problem, being sometimes insufficient to reduce environmental externalities in small economies (Balineau and de Melo 2011).

Environmentally oriented trade policies could motivate the implementation of technology innovation in developing countries to address CC policies; however, trade is not the only means for this purpose since foreign direct investment (FDI) plays a key role in the environmental and external dividends. Dechezleprêtre et al. (2009) highlight the role of technology transfers in GHG mitigation in Mexico, Brazil, China and India. Even though the patterns differ across countries from different continents, they conclude that the success of technology transfer for CC mitigation in developing countries is highly dependent on the foreign contribution (FDI such as projects in subsidiaries, imports), but also on the capacity building in the host countries to accelerate technology diffusion domestically. Forsyth (2007) also supports the need for technology transfer for CC mitigation in developing countries (India, the Philippines and Thailand) where establishing partnerships domestically (across sectors and public–private cooperation) and with foreign partners can reduce economic costs of delivering cleaner mechanisms for the development dividend.

Similarly to the definition of instruments, these trade-environmental agreements do not establish specific penalties in case of non-compliance with the commitments made. However, recent discussions in this regard present some trade policy instruments as a means of sanctioning environmental inaction (*e.g.* inaction under the Paris Agreement). The *embodied carbon tariff* (Böhringer et al. 2016) could be an example of trade-environmental policy instruments taxing trade based on carbon content. From the theoretical point of view, it turns out to be equitable to correct the environmental externality generated by international exchanges, but from the point of view of its implementation (the absence of complete information by country, by product and some organizational issues) are, for the moment, impracticable.

Beyond that, the threat or risk of an international sanction through trade policy is latent and could act as dissuasive measures for the adoption of national environmental measures.

In this context, which requires the coordination of environmental policies at the international level and where the urgency of climate phenomena is utmost, the threat of these possible sanctions could push Latin American countries to carry out ETRs. Of course, such ETRs must take into account the stylized facts and structural characteristics of developing countries mentioned above. Consequently, an ETR, such as the tax on GHG emissions, should come to replace distortionary taxes both from the functional and personal distribution of income, while maintaining/improving the capacity of the Treasury and private investment financing.

Summing up this review of the literature, there are two important fronts to appraise the costs of CC policies in developing economies, such as Argentina's. The first one is the **domestic front**, in terms of how wages are determined institutionally or by the market, because additional costs could be passed through to wages and therefore reduce employment (with the obvious political consequences). The second one is the **external front**: well-intended initiatives of individual countries could be jeopardized by the stress of the balance of payments, when foreign resources are necessary but costly; by loss of competitiveness when not accompanied by the rest of the world (since a simultaneous move of all countries could help to stabilize the relative changes in competitiveness), or by the migration of capital to other regions of the world, which could limit or reverse the expected gains of an ETR.

Argentina and the CC Commitments

The tax structure of developing economies, such as Argentina's, has historically been more in the positive than in the normative field, because the tax designs calculated from more traditional microeconomic methodology have had to give ground to macroeconomic imbalances, lobbies, exemptions for reasons of merit and income distribution, evasion and avoidance. Consequently, when thinking of an ETR to face CC issues, these structural particularities of Argentina have to be considered. This is the first step towards an appropriate ETR.

The second step is to identify the causes that generate carbon emissions to choose the right instrument to tackle both current and future emissions and even the stock due to past emissions. According to the statistics of the CAIT (2015), permanent flows in global GHG emissions are mainly linked to the (intermediate and final) consumption of fossil fuels, with the energy sector accounting for more than 70% of global emissions. The sub-sectors that use fossil energy mainly include electricity generation, heating and transportation globally. This is followed by the agricultural sector whose contribution to global emissions is between 14 and 11% (from 1990 to 2012) with a declining share since 2000.

The third step to think of an ETR for developing countries, such as Argentina, is to put in perspective the relative CC responsibilities of each country, so that the costly measures to be implemented become fair and equitable in the solution of this global environmental problem. In this regard, we will see that the Latin American countries have contributed little to the generation of the environmental liability and that they are currently small GHG emitters compared to the rest of the world.

The statistics from the CAIT 2015 Climate Data Explorer provide evidence that China, the USA and the European Union accounted for more than half of the global emissions in 2012, as well as a glimpse of the small contributions of large Latin American countries, such as Argentina (0.6%) and Brazil (1.4%), to them. The efforts to reduce carbon emissions by each of the countries must be comparable to the responsibility for the global damage that each of them causes, and, from this point of view, the mitigation efforts undertaken by the "environmentally" small countries are irrelevant if the big ones do not take the corresponding measures. If large countries contribute to solving the problem of their carbon emissions on the basis of their global damage, emissions from small countries could be more than compensated. Nevertheless, the reverse solution is not enough. In line with this description, good news comes from the ratification of the Paris Agreement, as more than 179 out of 197 countries have ratified the agreement through voting and the approval of their parliaments as at March 2019. Among them are the main emitters mentioned previously, but not all.

These global and national patterns of GHG emissions show a strong relationship with the economic activity, to which the role of the energy sector is central. Consequently, the investment in new renewable forms of energy and in the technological progress improvement could contribute to decoupling emissions from the economic growth (Kitous et al. 2016; Vandyck and Van Regemorter 2014). Some of these initiatives through the incentives provided by an appropriate ETR could lead to more than an environmental dividend for these environmentally small economies.

The challenge is difficult because the phenomenon in hand is new and not well documented. A scenario of pure uncertainty (in the Knightian sense) arises with respect to key parameters and causalities. Thus, in each case of study, it is necessary to address the right questions: How will the economy be affected by CC and by traditional policy instruments as part of the ETR? How are wages determined? Which is the degree of capital mobility with respect to the rest of the world? Could an environmentally small country reap more than one dividend of an ETR? Or could the adaption decision facing CC be preferred to mitigation policies for a country such as Argentina? Simulation models, and especially Computable General Equilibrium ones, seem to be an appropriate instrument to evaluate policies and actions concerning CC as a target (Wing 2004; Chisari et al. 2012).

Appropriate Tool for Measuring Dividends of CC Policies: A CGE Model for Argentina

Looking for the conditions under which multiple dividends of the climate change policy could arise in the case of Argentina, we use a CGE model based on Chisari and Miller (2015) calibrated to this country in 2006. The programming of this model as MPSGE makes it tractable to evaluate the key assumptions that constrain the emergence of positive side effects on the economy as a consequence of this environmental policy.

The following description of the CGE model assumptions is complemented by Annex A (equations of a simplified version of the model).

Main CGE Model Assumptions

We model the behaviour of a small open economy which faces fixed market prices for all tradable goods. The small economy assumption is true to Argentina in terms of both carbon emissions and EGS trade, which would be at the centre of our ETR evaluation. This assumption means that, compared to the rest of the world, Argentina is not big enough to impact on world prices of commodities; however, its ETR impacts on domestic prices and, thus, on its relative competitiveness in some markets. Moreover, given the sectoral aggregation of this multi-sector economy in 6 sectors, such as *Agriculture and Fishing* as an aggregate, the small open economy assumption remains realistic, since Argentina could not influence prices at that level of sector aggregation. Based on these six sectors, we identify Argentina's production/consumption patterns and the sectors' carbon intensity. The agents that interact in the model are these single-commodity firms, two types of households—rich and poor—, the government and the rest of the world. Below we describe the main assumption of agents' behaviour and closure of the model.

Firms. A representative firm of each sector maximizes its profits subject to the production function of a single commodity. The production function nests inputs and value added in the first level in a fixed proportion (Leontief's assumption). Then, two mechanisms that could reduce carbon emissions are modelled in the further nests of the production function. First, for inputs, we assume a two-level nest, where, in the first level, EGS and non-EGS inputs are combined according to a Cobb–Douglas assumption and then, only the EGS products show, in the second nest, a greater substitution between domestic and imported—assumed relatively cleaner—EGS products (Constant Elasticity of Substitution—CES—assumption). The second mechanism consists of reducing the carbon intensity of firms through the quality of the value added. The value-added nest is a Cobb–Douglas function that combines labour and capital. The capital could be domestic or foreign, considering the latter environmentally cleaner than the former. Thus, when foreign direct investment (FDI) is allowed, the production technology switched from a "dirty" (domestic capital) to a relatively "clean" (foreign capital) one in terms of its carbon emissions. These latent clean technologies are allowed in strategic sectors, such as Energy and Industry, but their implementation is not costless in terms of required exports. Figure 1 illustrates this production function tree that we have described.

Households. Households, rich and poor, consume domestic and imported goods and services, invest and buy/sell bonds in a constant proportion of their income (Cobb–Douglas assumption). Their incomes consist of labour and capital remunerations and of transfers received from the government and the rest of the world. Thus, each type of household maximizes its utility function subject to its budget constraints finding the optimal composition of its consumption basket of final goods and services. Like in the intermediate consumption of firms, the two-level nested final consumption demand of households assumes a first nest between EGS and non-EGS products (Cobb–Douglas assumption) and, then, a higher degree of substitution between domestic and imported EGS, since the latter are considered environmentally cleaner (CES). This characteristic of the final demand tree makes it possible to introduce a mechanism to green the economy through final consumption decisions. Figure 2 illustrates this demand function tree described above.

Government. The government also consumes, invests and makes transfers to households in a constant proportion (Cobb–Douglas), financing those expenses mainly with its tax collection (tariffs, labour and capital taxes, output taxes, carbon taxes) and debt in a lower proportion. In this sense, the modelling of the government behaviour is neutral because each dollar received by the government is always spent in the same way. Moreover, keeping the same level of revenues and expenses, the

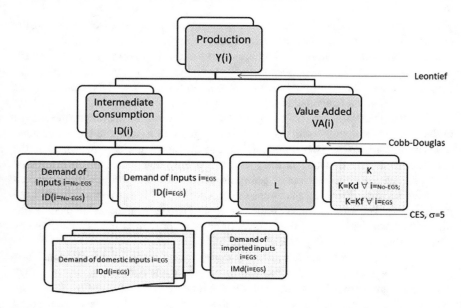

Fig. 1 Production function tree with input substitution (EGS and origins) and latent technologies (domestic, Kd versus foreign, Kf, capital)

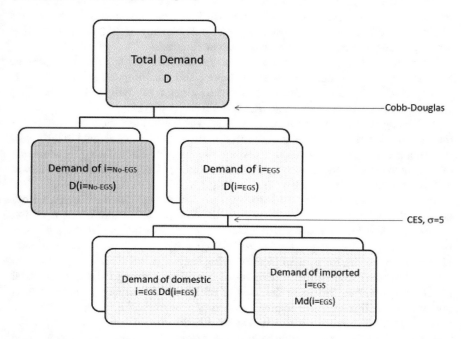

Fig. 2 Demand tree with substitution between EGS and non-EGS

government can make the decision to compensate taxes, for instance, increasing the carbon tax to reduce other distortionary taxes.

Rest of the World. The rest of the world buys domestic exports and sells imports in addition to making transactions in the financial market and collecting dividends from investment (FDI). In the benchmark situation, the value of exports equalizes the value of imports (trade balances).

The equivalent variation (EV) is the welfare measure chosen to evaluate the change in the level of the agents' utility when prices of goods, services and factors change.

Carbon Emissions. Two indicators of carbon emissions are measured at the national level. The first one is the total carbon emission index that is computed according to the consumption of polluting goods and the production of carbon-intensive sectors, taking into account the mechanism of latent technologies (fixed proportions). The second carbon indicator is our "Kuznets index", which compares the total carbon emissions with the level of activity, providing the carbon-intensity measurement of the country's GDP.

The net impact in terms of emissions depends on carbon inter-industrial transactions. For example, let us assume that an economy has only two goods (Q_1, Q_2), where each good contributes differently to the total carbon emissions (e_1 and e_2 are the coefficients of carbon emissions per unit of product Q_1 and Q_2, resp.); thus, total carbon emissions are: $CO_2E = e_1 Q_1 + e_2 Q_2$.

Then, the change in the total carbon emissions in an economy, when introducing, for example, an ETR on good 2 (*e.g.* t_2), will be broken down into three separate effects using the taxonomy proposed by Brock and Scott Taylor (2004). They identify three channels through which emissions can be reduced:

First, the *scale effect*, which takes into account how the scale of activity can change responding to taxes or other incentives, and thus, impacting on emissions. Technically, this effect is given by movements along a ray defined by $Q_2 = sQ_1$, where s is a positive number; then, $dCO_2E/dQ_1 = e_1 + e_2s$.

The second effect is the *composition effect*, which considers the modifications in the composition of the value added (in terms of the share of every activity in the total), reducing the relative participation of carbon-intensive activities. This effect depends on the movement of the economy along the frontier of possibilities of production $Q_2(Q_1)$, and thus, $dCO_2E/dQ_1 = e_1 + e_2Q_2'(Q_1)$.

Finally, the third effect is the *intensity or technique effect*, which derives from educing the coefficients of emissions per unit of output as a result of the adoption of alternative technologies, *i.e.* $dCO_2E/de_1 = Q_1$. This effect could be the result of substituting polluting production technologies by cleaner ones (less emission intensive). Allowing for a latent cleaner technology implementation (considered in this CGE model) leads to the abatement of emissions keeping the same industrial structure and without reducing the level of the economic activity. Nonetheless, as reviewed in the literature, the implementation of new green alternative technologies is not costless. The presence of sunk capital and the cost of opportunity of foreign resources could limit the introduction or the extent of application of a new clean technology that could change the intensity of emissions per unit of production. Thus, when the implementation of a cleaner production method becomes too expensive, an ETR will reduce emissions only through changes in the scale (reduction of GDP keeping the share of every industry constant) or through the change in the composition of value added (changes in the sectoral structure of GDP).

Closures. The model is tractable to introduce different closure assumptions for factor markets. In the case of the labour market, we assume three alternatives of closures: (i) full employment, (ii) positive unemployment rate (10.2%) due to constant wages in real terms; *i.e.* wages are indexed to the price of the consumption basket of a poor household, and (iii) positive unemployment due to fixed nominal wages in terms of the foreign currency. Depending on how wages are determined, institutionally or by the market, additional costs due to an ETR could be passed through to wages and therefore increase unemployment. In this sense, the costs of implementing an ETR to reduce carbon emissions will be greater under initial unemployment conditions than under full employment. Concerning the capital factor, the model assumes three types of capital: fixed or sector-specific capital, on one side, and on the other side, mobile domestic and foreign capitals, allocated across sectors according to their rates of return. Capital mobility is assumed to be initially low according to the characteristics of the Argentinean economy (*i.e.* only 12.5% of domestic capital of free mobility across sectors); but, given the flexibility of the model programming, a greater percentage of mobile capital (50%) could be implemented. When an ETR increases relative costs for polluting sectors, a greater capital mobility across sectors makes it possible to reduce economic costs (*i.e.* capital is reallocated to environmental cleaner sectors) and enables a better performance of the environmental policy. The tractability of the model is also applied to allow for (or not) the FDI, which is also a key assumption of the model for the implementation of an alternative cleaner production technology, which is not costless due to the presence of sunk capital and positive opportunity costs for foreign capital. We will also evaluate the consequences of an ETR under a higher degree of international capital mobility, allowing for the mechanism where higher capital mobility across countries would lead to a negative impact for a country's own economy with a potential carbon leakage problem. The foreign capital remuneration is considered as the *numeraire* of the model. Finally, in accordance with Walras' law, we find the equilibrium of the balance of payments by combining the solutions of the optimization problems of firms, households and the government in this economy and assuming market clearing conditions under perfect competition for goods and factors.

Dividends of an ETR. In order to evaluate the presence of multiple dividends of an ETR, we consider the following indicators associated with each dividend. Testing for the presence of a *double dividend*, we explore how the GDP and a carbon emission index respond to a change in the ETR (including both taxes on domestic goods and imports). The *third dividend* test inquires additionally whether the ETR reduces the rate of unemployment, poverty and income distribution indexes. And the *fourth dividend* also looks into the performance of exports. For instance, if exports have to be increased significantly as a consequence of a carbon tax, the environmental programme could fail. Even though most of the exported goods are commodities, a sudden increase in exports is unrealistic, and the lack of exports could put the foreign reserves of the economies under stress.

Table 1 Carbon emissions and trade protection by sector in Argentina (2006)

Sector	Description	EGS	% CO_2	tIC (%)	t_{FC} (%)	t_I (%)
S1	Agriculture and fishing	No	38	3.6	25.14	9.94
S2	Energy and mining	No	21	0.15	–	–
S3	Industry	Yes	12	–	28	14.94
S4	Electricity and water	Yes	3	–	–	–
S5	Transport	No	13	–	–	–
S6	Other services	Yes	12	–	23.2	–

Source Prepared by the authors

Notes t_{IC}, t_{FC}, t_I tariffs applied on intermediate consumption, final consumption and investment goods, respectively

As stated in the literature, the role of an ETR (*e.g.* carbon taxes) in approaching Pareto optimality depends on the initial tax structure of an economy. For example, a new *ad-valorem* charged on the final demand of one good could reduce losses due to distortions rather than increase them when the rest of the goods are already taxed.

Calibration Data

The CGE model is initially calibrated using Argentina's Social Accounting Matrices (SAM) for 2006. According to the sectoral disaggregation of the SAM, we work with six sectors: Agriculture and Fishing (S1), Energy and Mining (S2), Industry (S3), Electricity and Water (S4), Transport (S5) and Other Services (S6). Since the industrial sector is relatively less pollutant than agriculture and energy (Table 1), and since most EGS lists under discussion mainly concern manufactures, we assume S3, S4 and S6 as EGS. Imports of manufactured final consumption and investment goods are relatively highly protected in the baseline compared with agricultural goods. However, *ad-valorem* equivalent tariffs applied on energy and mining sectors are almost zero in the baseline. In this sense, trade liberalization scenarios on EGS will relatively change this trade protection pattern and, consequently, agents' decisions of demand. The same will be true when introducing a fixed amount taxing the carbon emissions of each sector, since, in our baseline, no carbon tax is initially calibrated.

Description of Climate Change Policy Scenarios

In order to add some empirical evidence concerning the conditions under which multiple dividends can arise from environmentally oriented policies applied by developing countries, we will simulate, in the case of Argentina, two main scenarios based on policy tools available for this type of countries.

The first scenario makes it possible to tax carbon emissions linked to energy-intensive sectors and production processes in a fixed amount of 20 USD per ton of carbon emissions. Polluting production processes (*e.g.* agriculture and cattle production) and energy consumption, as intermediate inputs as well as final goods, are taxed by this carbon tax. For instance, the energy consumed by the transport sector as input will be charged the *ad-valorem* equivalent 38% tax; and in the case of the agriculture and fishing sector, its energy intermediate consumption will be taxed up to 24% in *ad-valorem* equivalent plus a tax of 10.4% over its own production according to its carbon generation (for further details, see the annex of Chisari and Miller 2015). The introduction of this carbon tax will be simulated under two different situations in order to identify possible dividends of this ETR. First, we assume the application of this carbon tax without assuming any change in the current tax structure of the country. Then, in a second simulation, we compensate the reduction in distortionary taxes by the additional revenue due to the increase in the carbon tax. In order to look for a double (greater GDP) and third (lower unemployment and poverty) dividend, we reduce taxes on labour.

The second scenario concerning an ETR simulates the improvement in the EGS market access in Argentina as part of the plurilateral EGS negotiation (*i.e.* Environmental Goods Agreement—EGA).[4] According to the sector disaggregation of Argentina's SAM, we have assumed that industry, electricity and water, and other services are EGSs. This ETR assumes tariff reduction/elimination on EGS, capping them to 5% as a maximum tariff on EGS. According to the initial tariffs in Table 1, we eliminate those which are lower than 5% and we reduce to 5% those tariffs which are initially greater than 5%. Tariff reduction/elimination on EGS is applied on intermediate, final and investment goods. Tariffs on EGS are also eliminated/reduced by other EGA trade partners, leading to an increase in EGS world prices. The magnitude of the shock assumed was a 5% increase in EGS world prices. The purpose of this scenario is to induce changes in relative prices that incentivise the substitution from polluting to environmental goods for any of the aforementioned usages. This scenario appears in the context of the trade liberalization of EGS, where doubts emerge concerning the possibility of achieving the trade, development and the environment dividends simultaneously (WTO 2001). As discussed in previous literature, this goal is challenging for developing economies given their structural and macroeconomic constraints and, particularly, where the current state of technology is not adequate to mitigate climate change (Laborde and Lakatos 2012) and where the implementation of a cleaner technology could be costly (De Melo 2017).

Thus, the first ETR scenario directly targets carbon emissions and the second one indirectly introduces incentives to reduce the carbon emissions through a greener trade policy reform.

Other environmental policy scenarios could also be simulated, such as subsidies on renewable energy production, but they imply additional fiscal costs that a developing

---

[4]The countries currently involved in the negotiation of an EGA are Australia, China, Costa Rica, the European Union, Hong Kong, Iceland, Israel, Japan, Korea, New Zealand, Norway, Singapore, Switzerland, Chinese Taipei, Turkey and the USA.

economy may not necessarily afford. Consequently, we leave aside this option of an ETR in Argentina.

The literature highlights the possibility that multiple dividends may emerge as a consequence of an ETR that depends on factors market conditions. Looking for those multiple dividends, we will evaluate the sensitivity of the results of these scenarios to:

1. Different conditions for the labour market: full employment, unemployment due to real constant wages and unemployment due to nominal constant wages.
2. Different degrees of national and international capital mobility: low vs. high inter-sectoral capital mobility.
3. The possibility of implementing a more ecological production technology (modelling of latent technology) through the imports of capital goods and FDI in the EGS market opening scenario.

Multiple Dividends: Results and Discussion

We illustrate the results of applying two ETRs in the case of Argentina, a carbon tax and a trade agreement on EGS, in order to find the conditions mentioned in the literature for multiple dividends of an ETR: lower carbon emissions, GDP growth, lower unemployment rates and more equity (welfare increase for poor households), and trade increase without too much pressure on the external balance (real exports). These are the indicators we will look at for simultaneous dividends of the ETR.

We present the results as radar (or spider) charts in order to easily identify multiple dividends or possible trade-offs between the aforementioned indicators. The results of all indicators are presented in percentage variation compared to the baseline, except for unemployment, which is displayed in rates. The baseline scenario is thus set at the zero point for all indicators, but for the unemployment rate, the baseline level is 10%.

Carbon Tax and Its Potential Multiple Dividends

Let us start looking at Fig. 3, where the introduction of a carbon tax is unambiguously effective to reduce carbon emissions (-5.83%) in an economy with full employment, also leading to a slight increase in real exports (1.18%). However, costs in terms of real GDP (-0.1%) and income distribution (-0.3%) arise because of this ETR. According to these results, we can say that the reallocation of resources to relatively less polluting sectors (composition effect) will be greater than a lower scale to account for the reduction in carbon emissions. Nevertheless, this is not a realistic benchmark for a developing economy like that of Argentina, which has a structural positive unemployment rate (10% in the baseline). As previously discussed in the review of the literature concerning the double and the third dividends of an ETR, with the

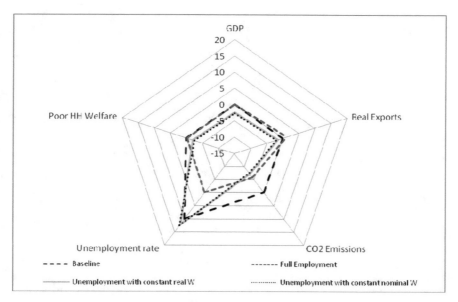

Fig. 3 Carbon Tax in Argentina under different functioning conditions of the labour market. *Source* Prepared by the authors based on Chisari and Miller (2015)

introduction of a carbon tax under unemployment (either due to nominal or real constant wages), the socio-economic costs of the ETR become greater and thus less applicable in the case of developing economies. This first comparison which takes into account the sensitivity of a carbon tax to labour market conditions is also illustrated in Fig. 3. With initial labour unemployment, a carbon tax leads to a significant GDP (between −2.4 and −2.8%) and export reduction (between −1.5 and −2%) due the loss in competitiveness, which is even greater compared to the full employment situation. Moreover, since the carbon tax increases costs of production and the level of activity falls, firms reduce their labour demand leading to an increase in the unemployment rate (from 10 to 12% and 12.5%). This result also leads to the deterioration of the income distribution between poor and rich households (between −2 and −2.5%). It is important to remark that previous negative impacts are greater when wages are fixed in terms of the foreign currency (nominal rigidity).

In order to look for multiple dividends, we evaluate the application of a carbon tax to replace the revenue of a distortionary tax on labour. Figure 4 illustrates this ETR with lower labour taxes, where no multiple dividends arise either under a full employment assumption or under unemployment due to nominal rigidities. Under full employment of labour supply, this compensation between a distortionary tax on labour and the carbon tax also leads to slight efficiency gains for the economy (almost the same results as in Fig. 3). And when wages are fixed in nominal terms, which means that they are fixed in terms of the *numeraire* (the remuneration of the

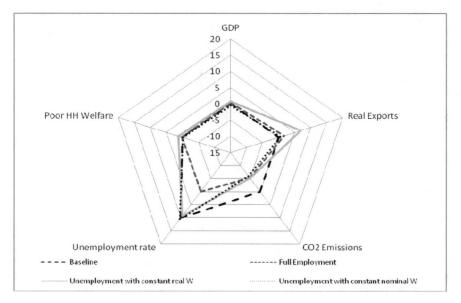

Fig. 4 Carbon tax in Argentina with domestic tax compensation. *Source* Prepared by the authors based on Chisari and Miller (2015)

foreign production factor), the socio-economic costs of this ETR become higher at the expense of the environmental dividend only.

Nonetheless, when unemployment is due to constant real wages (wages indexed to local prices), the lower tax charged on labour motivates a greater labour demand by local firms and thus impacts positively both on employment and GDP (Fig. 4). Therefore, multiple dividends of this ETR with a distortionary tax compensation emerge: the first dividend is lower carbon emissions (-5.12%), the second dividend is GDP growth (0.71%), the third dividend is lower unemployment rate (from 10 to 9.34%) and a better income distribution between poor and rich households (1.2%) and, finally, the fourth dividend is an increase in real exports (6.95%) because of an improvement in the international competitiveness. This pattern of multiple dividends of an ETR intensifies when the labour tax is reduced in compensation of higher carbon tax revenue (Chisari and Miller 2015).

The result of these multiple dividends is accounted for by the initial distortionary tax structure; thus, the replacement of those domestic taxes by a carbon tax makes it possible to reduce the environmental externality while gaining efficiency and equity.

In short, distortionary taxes impact more heavily on sectors that are labour intensive, mostly service sectors that are not intensive in terms of use of energy. When labour taxes are replaced using carbon taxes, there is a reallocation of resources and a change in GDP composition favouring service sectors and reducing carbon emissions. At the same time, the reduction of labour taxes provides an incentive to firms for hiring more workers; in turn, this becomes a benefit for workers and increases the welfare of the poor. The grossing-up reduction of labour costs represents a decrease in domestic costs vis-à-vis the international costs, and that is why we observe an improvement of the trade balance.

Developing countries, such as Argentina, have to pay attention to the degree of capital mobility across sectors and also to their capital mobility compared to the rest of the world. Figure 5 illustrates the application of a carbon tax, such as under Fig. 3, but when there is greater capital mobility across sectors and countries. In these situations, a more stringent ETR would reduce carbon emissions (between −13 and −11%) mainly due to a lower scale effect since the carbon tax increases the costs of production in a context where capital can outflow elsewhere. So, greater capital mobility across sectors and, particularly, compared to the rest of the world could eliminate any possibility of multi-dividends of a carbon tax. Once again, the carbon tax will be the first best option to reduce carbon emissions, but socio-economic costs will magnify.

Summing up the results of applying a carbon tax in a developing economy to comply with the international climate change commitments, we can say that multiple dividends can only be reached when local regulation manages low capital mobility in relation to the world, when wages rigidities are in terms of local purchasing power (not foreign) and, finally, when the government has the possibility and the intention to reduce the inefficiencies of its tax structure by compensating distortionary taxes, such as labour taxes, with the carbon one.

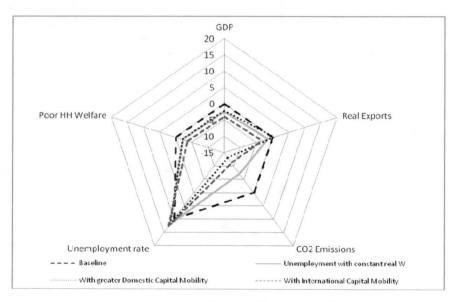

Fig. 5 Carbon tax in Argentina under different capital mobility assumptions. *Source* Prepared by the authors based on Chisari and Miller (2015)

Trade Liberalization on EGS as Second-Best ETR

Even though the carbon tax appears as the first best ETR to reduce carbon emissions, we have realized that the implementation of multiple dividends for developing economies is constrained to the presence of inefficiencies in the tax structure, particular factors markets conditions and the government's possibility for tax compensation (not higher fiscal deficit).

So, some other options appear as ETRs, for instance, through tariff reforms. Given the discussion in the multilateral arena (the Doha Round), by the APEC and currently in a plurilateral way by a small group of countries, concerning the incentives for trade in EGS, we evaluate the consequences of an EGA in the case of Argentina. This scenario could be an alternative to the carbon tax for developing countries to indirectly tackle climate change commitments, while looking for other socioeconomic dividends.

The change in the local and foreign tariffs favouring EGS stimulates real exports and GDP and reduces the unemployment rate and income disparities between poor and rich households. However, total carbon emissions do not fall in absolute terms, and even increase slightly. This is the common pattern of results displayed by Fig. 6 under an unemployment assumption due to constant real wages, with and without latent cleaner production technologies and different (low/high) degrees of capital mobility across sectors.

In this sense, it is possible to think that an EGS-oriented trade liberalization is not really effective for this environmental objective. Nonetheless, it is necessary to

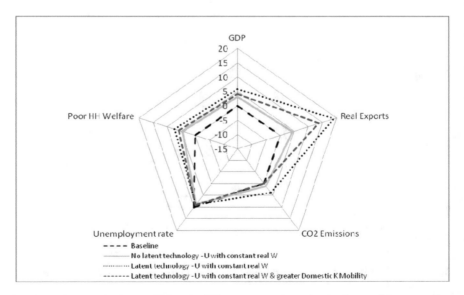

Fig. 6 Trade liberalization on EGS under different assumptions of technology and factors market. *Source* Prepared by the authors based on Ramos et al. (2017)

analyse the change in carbon intensity of GDP and trade, and eventually to break down the total change in CO_2 emissions in the three effects (scale, composition and technique) for a better understanding of this ETR.

Thus, first comparing the CO_2 emission change rate with the GDP growth rate, we can say that the carbon intensity of GDP has fallen even when total CO_2 emissions slightly increase under the EGA. This result is preferred to the reduction in total CO_2 emissions at the expense of GDP, trade or employment costs as observed when applying a carbon tax under unemployment conditions (also assumed here) without any tax compensation (Fig. 3).

Figure 6 makes it possible to analyse the EGS trade liberalization scenario in terms of carbon intensity and socio-economic dividends. Pointing out the different technological assumptions, we can see that when the use of latent greener technologies is allowed, the impact on the socio-economic dividends magnifies. The change in technology to a less polluting one makes it possible to identify a technique effect (-2.24%), which, added to the composition effect (-2.51%) induced by the change in relative prices between EGS- and non-EGS products, counterbalances almost 60% of the scale effect (8.58%) due to a greater level of activity.

Like in the carbon tax scenarios, the degree of capital mobility across sectors also impacts on the results. Greater inter-sectoral capital mobility combined with latent ecological technologies retains the increase in all indicators (even in total CO_2 emissions); however, the relation between CO_2 emission and GDP growth rates still favours GDP leading to a less carbon-intensive economy. This effect is due to the easy capital reallocation to the EGS sectors which also require foreign capital when installing the greener technology. In this case, the addition of the composition and technique effects (-5.51% and -1.45%, respectively), which reduces emissions, compensates almost the entire scale effect (6.65%) that increases them.

Even though the scenarios with latent technologies are preferable in terms of lower carbon intensity (Fig. 6), the increase in exports reveals an important pressure on the external balance to finance the foreign new technology (19% increase in real exports). This pressure is reduced when allowing for a freer allocation of capital across sectors but remains high anyway (13.7% increase in exports). This external result could be the real constraint for the signature of an EGA by countries like Argentina.

In short, we can say that a trade reform based on an environmental purpose, such as the EGA, could appear as an ETR option for developing economies to maximize socio-economic dividends while retaining the carbon emission increase, only if the need for exports to finance the imported technology remains reasonable.

Main Findings and Final Remarks

There is no doubt that we have to fight climate change. We have the alternative of reducing emissions or protecting our welfare with more adaptation to its consequences. But the cost of reducing emissions reveals itself as higher than expected, and higher than estimated in simulation models that assume full employment and limited mobility of capital. If the problem were cheap to solve, it would have already been solved.

This chapter has summarized the estimated costs when two main alternatives are considered: taxes on carbon emissions or relatively higher tariffs on trade that are not environmentally friendly, and we have seen that the costs are much greater under unemployment, high mobility of capital across countries and the adoption of costly technology.

Moreover, we have seen that some critical dimensions have to be taken into account when the special characteristics of less developed economies are considered. This creates a multi-dimensional problem with possibly critical associated thresholds. For instance, a policy aimed at reducing emissions that significantly increases unemployment will not be feasible, or a technological substitution that is very demanding in terms of foreign resources could be blocked by financial needs and trade balance results.

Annexes A

A simplified CGE model to estimate the costs and dividends of an ETR

In this annex based on Chisari and Miller (2015), we present the basic assumptions of the general equilibrium model used for simulations of the ETRs to Argentina. Changes in relative prices for goods (domestic and foreign) and factors (labour and capital) due to the implementation of an ETR, would lead to possible multiple dividends depending on the markets' behaviour constraints.

Even if the CGE model assumes more than one household (at least poor and rich), for simplification in notation let us assume that we have only one representative household that maximizes utility. Equation (1) gives the equalization of the subjective rate of substitution for an index of utility U with relative prices, corrected by *ad-valorem* taxes, in this case initially charged on good 1 (the general model includes several taxes, as well as agents and goods).

$$U_1/U_2 = (1 + t_1)P_1/P_2 \tag{1}$$

Equation (2) provides the budget constraint. It is assumed that there is only one kind of labour, whose endowment is denoted by L_0 (W is the wage rate) but two kinds of capital—fixed and mobile—between industries. There is one unit of specific capital in each industry, and it prices are indicated with π_i (alternatively, this can be interpreted as total profits of the sector i, with $i = \{1, 2\}$ with constant returns to scale).

The endowment of internationally mobile capital, owned by the domestic household, is given by K_0 and its remuneration is R^*. At the benchmark the proportion of fixed capital owned by the domestic household with respect to mobile capital is therefore $2/K_0$ (in fact, this parameter can be unobservable and uncertain).

$$P_1 C_1 (1 + t_1) + P_2 C_2 = W L_0 + R^* K_0 + 1\pi_1 + 1\pi_2 \tag{2}$$

Equations (3)–(6) give the definition of profits for sector 1, the production function, and the optimal benefits first-order conditions, respectively. The price received by producers is net of expenses in intermediate inputs, both domestic and imported (given by parameter a and α). Imported goods are used as the *numeraire,* and the tariff applied on them is denoted by t_M. This tariff rate t_M is initially the same for any imported good, but under the scenario of EGS trade liberalization it will be eliminated on the relatively clean product, inducing the change in relative prices against the polluting one. Equations (7)–(10) are the analogous equations for sector 2.

$$\pi_1 = (P_1 - P_2 a - \alpha(1 + t_M))Q_1 - W L_1 - R^* K_1 \tag{3}$$

$$Q_1 = F(L_1, 1, K_1) \tag{4}$$

$$(P_1 - a P_2 - \alpha(1 + t_M))F_L = W \tag{5}$$

$$(P_1 - a P_2 - \alpha(1 + t_M))F_K = R^* \tag{6}$$

$$\pi_2 = (P_2 - P_1 b - \beta(1 + t_M))Q_2 - W L_2 - R^* K_2 \tag{7}$$

$$Q_2 = G(L_2, 1, K_2) \tag{8}$$

$$(P_2 - P_1 b - \beta(1 + t_M))G_L = W \tag{9}$$

$$(P_2 - P_1 b - \beta(1 + t_M))G_K = R^* \tag{10}$$

Equation (11) represents the budget condition for the public sector; in this simplified case, it is assumed that all revenue is used to hire labour (L_g); however, in the general model, it also includes purchase of goods/services, transfers to households, investments and net changes in the financial result.

$$W L_g = t_1 P_1 C_1 + \alpha t_M F + \beta t_M G \tag{11}$$

Equations (12)–(15) are the equilibrium market conditions. The first includes exports, X; the third determines unemployment, Un, and the last gives the equalization of demand and supply of mobile capital.

$$C_1 + bQ_2 + X = Q_1 \tag{12}$$

$$C_2 + aQ_1 = Q_2 \tag{13}$$

$$L_1 + L_2 + L_g + Un = L_0 \tag{14}$$

$$K_1 + K_2 + K_m = K_0 \tag{15}$$

Equation (16) fixes the price of good 1 at the level given by the rest of the world because it is a tradable good. This assumption of a small economy will be also considered for other tradable goods in the CGE model used for simulations.

$$P_1 = P^* \tag{16}$$

Equation (17) represents nominal wages determination as a weighted average of prices of tradable goods, non-tradable goods and imports (it is assumed that the price of imports is 1 but if the tariff is positive we have to consider t_M). Even under wages indexation, this equation becomes not operative in a dynamic model when capital accumulation increases faster than population, since all unemployment will be absorbed.

$$W = \gamma_1 P_1(1 + t_1) + \gamma_2 P_2 + \gamma_3(1 + t_M) \tag{17}$$

In Eq. (18), we define imports, M, limited to those for industrial uses, which in this simplified version does not include imports of final goods; however, in the CGE model this definition also includes imports of final and intermediate goods.

$$\alpha Q_1 + \beta Q_2 = M \tag{18}$$

In this simplified version of the model the net result in terms of carbon emissions, CO_2E, depends on carbon inter-industrial transactions; however, in the computed model, it also takes into account final goods transactions. For example, let us assume that total carbon emissions can be written as:

$$CO_2 E = e_1 Q_1 + e_2 Q_2 \tag{19}$$

where the coefficients e_i stand for the carbon emissions per unit of total product. Moreover, such as it was described in the subsection 2.1 the impact on the total

CO_2E could be decomposed in scale, composition and intensity effects, and we also considered them in the CGE model.

The 19 unknowns are: $P_1C_1P_2C_2W\pi_1\pi_2L_1L_2UnK_1K_2Q_1Q_2L_gMXK_mCO_2E$.

Given this simplified model we will consider a *double dividend* when the ETR allows reducing CO_2E while GDP increases or at least does not fall. A *third dividend* of the ETR will be added when the unemployment rate, Un, also falls and even a poverty indicator, such as welfare of the poorest households improves. Finally, the *fourth dividend* of the ETR would arise when exports, X, are not required to significantly increase as consequence of its implementation (e.g. to introduce a foreign cleaner new technology of production); otherwise, the ETR could fail. Even though most of exported goods are commodities a sudden increase of exports is unrealistic and thus, the lack of exports could put under stress the foreign reserves of the economies, which is a recurrent problem in developing countries.

As evidenced by the literature, not only factor market assumptions (e.g. wages determination and capital mobility) but also the initial tax structure of the economy, could condition the role of carbon taxes in approaching Pareto optimality. For instance, in this simplified model, a new *ad-valorem* tax (e.g. carbon tax) t_2 charged on final demand for the good 2 could reduce losses due to distortions rather than increase them (when $t_2 = t_1$).[5] See the rest of the model explanation with ETR in subsection 2.1.

References

Aldy JE, Levy E, Parry I (2010) What is the role of carbon taxes in climate change mitigation?" PREMnotes 2. Washington, DC, United States: World Bank. Available at: http://www1.worldb ank.org/prem/PREMNotes/Note2_role_carbon_taxes.pdf

Allan G, Lecca P, McGregor P, Swales K (2014) The economic and environmental impact of a carbon tax for Scotland: a computable general equilibrium analysis. Ecol Econ 100:40–50

Balineau G, de Melo J (2011) Stalemate at the negotiations on environmental goods and services at the Doha Round. FERDI Document de Travail (28)

Bento AM, Jacobsen M (2007) Ricardian rents, environmental policy and the 'double-dividend' hypothesis. J Environ Econ Manage 53(1):17–31

Böhringer C, Carbone J, Rutherford TF (2016) Embodied carbon tariffs. Scandinavian J Econ

Bovenberg AL, Goulder LH (2002) Environmental taxation and regulation. In: Handbook of public economics, vol 3, pp 1471–1545. Elsevier

Bovenberg AL, Van der Ploeg F (1994) Consequences of environmental tax reform for involuntary unemployment and welfare. CentER Discussion Paper 1994

Brock W, Scott Taylor M (2004) Economic growth and the environment: a review of theory and empirics. NBER Working Paper 10854. National Bureau of Economic Research, Cambridge, United States

CAIT Climate Data Explorer (2015) World Resources Institute, Washington, DC. Available online at: https://cait.wri.org

Carraro C, Galeotti M, Gallo M (1996) Environmental taxation and unemployment: some evidence on the 'double dividend hypothesis' in Europe. J Public Econ 62(1–2):141–181

---

[5]This would be a case of "double dividend" in the weak sense in terms of Zhang y Baranzini (2000).

Castiglione C, Infante D, Smirnova J (2018) Non-trivial factors as determinants of the environmental taxation revenues in 27 EU countries. Economies 6(1):7.S

Chisari OO, Galiani S, Miller S (2016) Optimal climate change adaptation and mitigation expenditures in environmentally small economies. Economia (LACEA) 17(n. 1)

Chisari OO, Miller S, Maquieyra JA (2012) Manual sobre Modelos de Equilibrio General Computado para Economías de LAC con Énfasis en el Análisis Económico del Cambio Climático. IDB-TN 445. Países de América Latina y el Caribe con aplicaciones a Cambio Climático", Washington, DC, United States. IDB-TN-445, Washington D.C

Chisari OO, Miller SJ (2015) CGE modeling: the relevance of alternative structural specifications for the evaluation of carbon taxes' impact and for the integrated assessment of climate change effects: simulations for economies of Latin America and the Caribbean. Inter-American Development Bank

Ciaschini M, Pretaroli R, Severini F, Socci C (2012) Regional double dividend from environmental tax reform: an application for the Italian economy. Res Econ 66(3):273–283

Dechezleprêtre A, Glachant M, Ménière Y (2009) Technology transfer by CDM projects: a comparison of Brazil, China, India and Mexico. Energy Policy 37(2):703–711

De Melo J (2017) Moving on towards a workable climate regime. In: Sustainable growth in the EU, pp 231–256. Springer, Cham

de Mooij RA (ed) (2000) Environmental taxation and the double dividend. Emerald Group Publishing Limited

Fan JL, Liang QM, Wang Q, Zhang X, Wei YM (2015) Will export rebate policy be effective for CO_2 emissions reduction in China? A CEEPA-based analysis. J Clean Prod 103:120–129

Fontagné L, Fouré J (November, 2016a) Changement climatique et commerce: quelques simulations de politique économique. Premier Ministre

Fontagné L, Fouré J (2016b) Long term socio-economic scenarios for representative concentration pathways defining alternative CO_2 emission trajectories. CEPII Research Report 2016-1. Available at: http://www.cepii.fr/CEPII/fr/publications/rr/abstract.asp?NoDoc=8565

Forsyth T (2007) Promoting the "development dividend" of climate technology transfer: can cross-sector partnerships help? World Dev 35(10):1684–1698

Fraser I, Waschik R (2013) The double dividend hypothesis in a CGE model: Specific factors and the carbon base. Energy Econ 39:283–295

Freire-González J (2018) Environmental taxation and the double dividend hypothesis in CGE modeling literature: a critical review. J Policy Model 40:194–223

Giménez EL, Rodríguez M (2010) Reevaluating the first and the second dividends of environmental tax reforms. Energy Policy 38(11):6654–6661

Gozlan E, Ramos MP (20–22 November, 2008) Trade liberalization in environmentally preferable products. Paper presented at the 2008 LACEA/LAMES joint conference. Rio de Janeiro

Grottera C, Pereira AO Jr, La Rovere EL (2017) Impacts of carbon pricing on income inequality in Brazil. Climate Dev 9(1):80–93

Karydas C, Zhang L (2017) Green tax reform, endogenous innovation and the growth dividend. J Environ Econ Manage

Kitous A, Keramidas K, Vandyck T, Saveyn B (2016) Global energy and climate outlook (GECO 2016) road from Paris (No. JRC101899). Joint Research Centre (Seville site)

Klenert D, Schwerhoff G, Edenhofer O, Mattauch L (2016) Environmental taxation, inequality and Engel's law: the double dividend of redistribution. Environ Resource Econ 1–20

Kolsuz G, Yeldan AE (2017) Economics of climate change and green employment: a general equilibrium investigation for Turkey. Renew Sustain Energy Rev 70:1240–1250

Koskela E, Schöb R (1999) Alleviating unemployment: the case for green tax reforms. Eur Econ Rev 43(9):1723–1746

Koskela E, Schöb R (2001) Optimal factor income taxation in the presence of unemployment. Discussion Paper 758. Research Institute of the Finnish Economy, Helsinki, Finland

Kuralbayeva K (2015) Environmental policy and the triple dividend in developing economies with rural-urban migration, mimeo

Kuralbayeva K (2018) Unemployment, rural–urban migration and environmental regulation. Rev Develop Econ 22(2):507–539

Laborde D, Lakatos C (2012) Market access opportunities for ACP countries in environmental goods. International Centre for Trade and Sustainable Development (ICTSD), Genève. Programme on Trade and Environment, Issue paper no. 17

Layard R, Nickell S, Jackman R (1991) Unemployment: macroeconomic performance and the labour market. Oxford University Press, Oxford

Marchand M, Pestieau P, Wibaut S (1989) Optimal commodity taxation and tax reform under unemployment. Scandinavian J Econ 91(3):547–563

Markandya A, González-Eguino M, Escapa M (2013) From shadow to green: linking environmental fiscal reforms and the informal economy. Energy Econ 40:S108–S118

Schöb R (March, 2003) The double dividend hypothesis of environmental taxes: a survey, mimeo

Tol R (2008) Why worry about climate change? A research agenda. Environ Values 17:437–470

Ramos MP (2014) The impact of trade liberalization of environmental products on welfare, trade, and the environment in Argentina. UNCTAD Virtual Institute Project for Trade and Poverty, Tech. Rep. Available at: https://vi.unctad.org/tap/docs/other/argentina.pdf

Ramos MP, Chisari OO, Martínez JPV (2017) Scale, technique and composition effects of CO_2 emissions under trade liberalization of EGS: a CGE evaluation for Argentina, world academy of science, engineering and technology. Int J Soc Behav Edu Econ Business Ind Eng 11(7):1744–1748

Rivera GL, Reynès F, Cortes II, Bellocq FX, Grazi F (2016) Towards a low carbon growth in Mexico: is a double dividend possible? A dynamic general equilibrium assessment. Energy Policy 96:314–327

Vandyck T, Van Regemorter D (2014) Distributional and regional economic impact of energy taxes in Belgium. Energy Policy 72:190–203

Van Heerden J, Gerlagh R, Blignaut J, Horridge M, Hess S, Mabugu R, Mabugu M (2006) Searching for triple dividends in South Africa: fighting CO_2 pollution and poverty while promoting growth. Energy J 27(2):113–141. Retrieved from http://www.jstor.org/stable/23297022

Wing IS (2004) Computable general equilibrium models for the analysis of energy and climate policies. Int Handbook Energy Econ

Carbon Taxes and Renewable Energy Subsidies: A Discussion About the Green Paradox

Maria Elisa Belfiori

Abstract What is the optimal policy to control carbon emissions that cause climate change? Climate change is a global phenomenon, and the solution to it is a carbon tax. What is an optimal policy? An optimal policy tackles the problem with the least economic distortions. Under some assumptions, the optimal carbon tax depends on three parameters only, which is very useful because it provides a simple rule for governments to follow. A related strategy, broadly adopted in many countries worldwide, is the subsidization of renewable energy. Subsidies to renewable energy are optimal if there are externalities in the sector. These may come from a learning-by-doing curve or technological spillovers. Otherwise, the subsidies add undesired economic distortions. Furthermore, renewable subsidies can be counterproductive as they may lead to an increase in overall emissions.

Keywords Climate change · Carbon tax · Renewable subsidies · Optimal policy · Green paradox

In Need: A Global Climate Policy

Carbon emissions are global and long-lived. About 80% of carbon dioxide released into the atmosphere has a mean lifetime of 300 years (Archer 2005), and this is regardless of where emissions originate. From an economic perspective, climate change is the problem of an externality. Carbon emissions released while burning fossil fuels contribute to global warming, and nobody pays for its consequences.

M. E. Belfiori (✉)
School of Business, Universidad Torcuato Di Tella, Av. Figueroa Alcorta 7350, C1428BIJ Buenos Aires, Argentina
e-mail: ebelfiori@utdt.edu

© Springer Nature Switzerland AG 2021
M. E. Belfiori and M. J. Rabassa, (eds.) *The Economics of Climate Change in Argentina*, The Latin American Studies Book Series,
https://doi.org/10.1007/978-3-030-62252-7_6

115

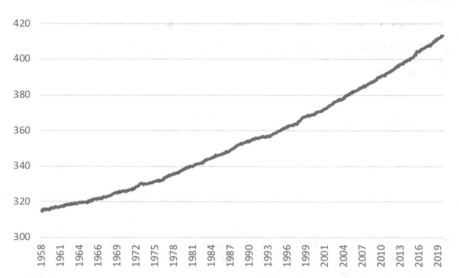

Fig. 1 Carbon dioxide (in parts per million) *Source* National Oceanic and Atmospheric Administration, US Department of Commerce. Carbon dioxide measured in Mauna Loa, Hawaii. Monthly data in parts per million

Carbon emissions must be priced, so polluters will have to pay for the damages caused by their emissions.

The solution to the climate problem is a global carbon tax, where "global" means that the tax rate must be the same across countries. The uniformity and coordination among countries are crucial. According to (WorldBank and Ecofys 2018), there are 25 emissions trading schemes and 26 carbon taxes worldwide. Some are implemented at national level and some, at sub-national level.

However, these carbon control policies cover about 20% of global emissions. Figure 1 shows global CO_2 emissions in parts per million from 2006 to today. The reading is clear: Carbon emissions are rising. The well-intentioned, but isolated, efforts to control carbon emissions have not been successful in cutting down emissions, not even to slow them down. Governments around the globe are relying on a wide range of climate policies with the hope of moving in the right direction. Still, politicians must coordinate and implement the optimal policy to fully fix the climate problem.

This chapter describes a framework to design an optimal climate policy and discusses some of the main findings in existing literature. An optimal policy achieves a given goal by maximizing social welfare. That way, no one can be better off unless someone else is worse off. This criterion corresponds to optimality in the Pareto sense or a first-best. Anything else is a second-best. Importantly, when governments implement second-best policies, some people are better off, and others are worse off. There is no clear way of ranking the possible outcomes. Distortions arise, and there is the risk of not achieving the goal.

The goal is not merely to cut emissions to zero. Emissions come from the production of consumption goods, and consumption is good for the economy. Societies that enjoy higher consumption are happier and wealthier. Thus, the optimal policy is to pollute at the optimal level that maximizes social well-being.

Integrated Assessment Models

It is useful to consider a global economy to think about this problem. Economics has developed a toolbox to study how the economy and the climate interact and, in particular, what the optimal carbon emissions tax is. Nordhaus and Boyer (2003) pioneered this research agenda, and there is now a burgeoning literature on this field (see Golosov et al. 2014; Acemoglu et al. 2012; Stern 2008; Barrage 2018, among others).

The building block in the analysis is an integrated assessment model. The name reflects the blend between an economic model and a climate model at the intersection of economics, physics and chemistry. Typically, the economic model is a version of a neoclassical growth model extended to include fossil fuels, a climate externality associated with the use of fossil fuels and a carbon cycle that determines how carbon in the atmosphere evolves and translates into a global temperature increase.

In the simplest of these models, a single consumption good is produced with renewable and non-renewable energy. The non-renewable energy source is oil. World oil reserves are draining as oil is used up for production. Renewable energy, such as solar or wind, is clean. Its technology relies on labor only.

There is a climate system embedded in the model to address climate change. The first component of this system is atmospheric carbon. Carbon increases with oil extraction. Scientific research reports that emissions remain in the atmosphere for a long time, and there is a low rate of natural carbon depreciation or reabsorption.

Twenty per cent of the emissions stay in the atmosphere forever. The remaining eighty per cent of these emissions has an average life of 300 years (Archer 2005). This means that, even though there is a natural reabsorption of human-induced pollution, it happens at a meager rate.

The last element that completes an integrated assessment model has to do with the presence of carbon in the atmosphere and how it becomes a cost to the economy. There are two common ways of adding this cost: as an output loss or a welfare loss. Nordhaus estimates that if the economy reaches a global temperature increase of 2.5–3 °C above the average temperature in the pre-industrial era, the world output loss will amount to approximately 0.48%. If global temperature increases further, the losses could be much higher due to the nonlinearities in the system that lead to catastrophic economic scenarios.

At this point, it is convenient to set up a model to investigate further on these issues. The non-technical reader may wish to skim through the technical details that follow.

A Simple Climate Change Model

Consider the following economy. Time is discrete and infinite, $t \in \{0, \ldots, \infty\}$. A continuum of identical individuals populates the economy. There are three production units: the final good and the energy producers, indexed by $i = \{0, 1, 2\}$. The consumption good is produced with capital K, labor N and energy E according to the following production function

$$\tilde{F}(A_{0t}, N_{0t}, K_{0t}, E_{0t}) \tag{1}$$

where A is sectorial productivity and E_{0t} is an energy composite. The stock of capital $\{K_t\}_{t=0}^{\infty}$ is exogenously given. Hence, there is no capital investment decision in this model.

Energy comes from two sources, an exhaustible resource (E_1) and green energy (E_2) according to

$$E_t = [\kappa E_{1t}^{\rho} + (1 - \kappa) E_{2t}^{\rho}]^{1/\rho} \tag{2}$$

where the parameter ρ represents the elasticity of substitution between the energy components.

The exhaustible resource is costless to extract. It can be thought of as oil or natural gas. At each point in time, oil use equals total oil extraction from oil reserves R_t

$$E_{1t} = R_t - R_{t+1} \tag{3}$$

The economy starts with an initial stock of oil, R_0. Renewable energy production uses capital and labor and has sector-specific productivity given by A_{2t}. Thus,

$$E_{2t} = f(A_{2t}, N_{2t}, K_{2t}) \tag{4}$$

The function f exhibits constant returns to scale and satisfies the Inada conditions.

Oil increases carbon in the atmosphere, S_t. In particular, carbon in atmosphere in every period t evolves according to

$$S_{t+1} = (1 - \gamma)S_t + E_{1t} \tag{5}$$

where $\gamma \in [0, 1)$ is the natural rate of carbon reabsorption, and the economy starts with a stock of carbon S_0.

The stock of carbon in the atmosphere generates a climate externality that takes the form of an output loss. Thus, total output is given by

$$Y_t = F(S_t, A_{0t}, N_{0t}, K_{0t}, E_{0t}) = [1 - x(S_t)]\tilde{F}(A_{0t}, N_{0t}, K_{0t}, E_{0t}) \tag{6}$$

The damage function x is increasing, concave and twice differentiable with $\lim_{S \to \bar{S}} x'$ $(S) = 0$, where $\bar{S}$ represents a lower bound on the atmospheric CO_2 concentration. The amount of labor is exogenously given and can vary over time.

Individuals derive utility from consumption and leisure and discount the future with the discount factor $\beta \in (0, 1)$. Over time, individuals care about the value

$$\sum_{t=0}^{\infty} \beta^t [u(C_t) - v(N_t)]. \tag{7}$$

The utility function u is increasing, concave and twice differentiable with $\lim_{c \to 0} u'$ $(c) = \infty$. The function v is increasing, convex and twice differentiable with $\lim_{N \to 0} v'$ $(N) = 0$. Individuals consume and work.

The feasibility constraints in this economy are given by

$$C_t + K_{t+1} = Y_t \tag{8}$$

for every period t, together with

$$N_t = \sum_{i=0}^{2} N_{it} \tag{9}$$

$$K_t = \sum_{i=0}^{2} K_{it} \tag{10}$$

for every period t.

The Best Outcome

What is the best path to follow to solve the climate problem? Economics calls it "the social optimum". The social optimum is the level of output, consumption, energy use and carbon dioxide that maximizes society's welfare. In this model, consumption increases social well-being, and work effort decreases it. In this sense, there exists a climate problem because consumption comes from oil extraction, which is polluting.

The best for society is not to cut emissions to zero—it can eventually be if countries do not implement climate policies soon enough. Instead, the best for society is to pollute at the optimal level: the level of carbon emissions that maximizes social well-being. The restriction is that global consumption cannot exceed global production, after taking into account the output loss from the climate externality.

Formally, the social optimum is the path $\{C_t, N_t, E_t, S_t\}_{t=0}^{\infty}$ that maximizes (7) subject to conditions (1)–(5) and (8)–(10).

The central element in the social optimum is the social cost of carbon. The social cost of carbon is the cost that climate change imposes on individuals and the economy. Formally, it is given by the following expression

$$\mu_t^* = \frac{\sum_{j=0}^{\infty} [\beta (1 - \delta)]^j u'(C_{t+j}^*) x'(S_{t+j}^*) F_{t+j}}{u'(C_t^*)} \tag{11}$$

Equation (11) states that the social cost of climate change is equal to the present value of the output losses associated with a marginal increase in carbon emissions. Emissions remain in the atmosphere for 300 years, on average, and some are everlasting (Archer 2005). Thus, damages will happen over time, and the social cost of climate change is equal to the present value of the future losses.

Efficient Energy Use

Another critical question concerns the efficient use of alternative energy sources. How much should the economy rely on renewable versus non-renewable energy?

There is a simple rule that governs the optimal use of non-renewable energy. This is the Hotelling rule (Hotelling 1931), and it is implied in Eq. (12). It says that the marginal benefit from oil extraction must be the same at any moment in time. If benefits are high in the future, it is better to keep oil underground and extract it later. The opposite holds if future benefits are low. In this case, it is better to extract more oil today. Therefore, an acceleration in oil extraction may be observed if oil companies foresee a drop in future benefits.

The benefit of oil extraction is that oil allows the production of consumption goods. This is the first term on the right-hand side of Eq. (12). Also, carbon emissions arise from oil extraction, and these costs must be net out from the benefits.

$$\beta u'(C_{t+1})[F'_{e_1,t+1} - \mu_{t+1}^*] = u'(C_t)[F'_{e_1,t} - \mu_t^*] \tag{12}$$

On the other hand, if there is a deferment of oil extraction, higher consumption will result in the future, and there will also be a delay in the occurrence of climate damages. This is the left-hand side of Eq. (12). Overall, the Hotelling rule solves the optimal depletion rate of global oil reserves.

The efficient use of renewable energy requires that the marginal rate of substitution between consumption and leisure (and work effort) equals the marginal rate of transformation. It is the standard condition for economic efficiency. It balances out the economic benefits and costs of producing renewable energy. Also, the efficient use of resources requires that the marginal productivity of factor inputs is the same across productive sectors. For renewables, in particular, this implies that

$$F'_{e_2,t} f'_{n,t} = F'_{n,t} \tag{13}$$

where the left-hand side of Eq. (13) is the labor productivity in the renewable sector and the right-hand side is the productivity in the consumption good production.

The Design of Climate Policies

Learning the features of the best outcome is essential for the design of optimal policies. It describes the efficient use of alternative energy sources and reveals whether there should be a transition to renewables. Perhaps, a mixture of different energy sources is desirable.

The optimal design of climate policies emerges from comparing the market outcome—that we observe in the real economy—with the best outcome.

The Market Outcome

This section lays out an economic environment equivalent to the environment described in Section "The Best Outcome", but from the perspective of a market economy. In a market economy, oil reserves belong to private companies. There are oil firms and renewable energy firms. Private firms make decisions to maximize their profits. This profit obviously depends on the market price.

If there is a carbon tax, the profit will be net of the tax paid. For renewable energy firms, the profit will depend on the price of renewable energy. Similarly, if there is a subsidy, then the profit will include the subsidy.

Formally, we can think of a market economy where there are three production units or "sectors" indexed by $j = 0, 1, 2$. Sector 0 corresponds to the final consumption good sector, and the remaining two are the energy producers.

A representative firm operates the technology (6) that produces the final consumption good. The firm hires labor at a wage w_t, rents capital from households at rate r_t and buys energy inputs from the energy sectors at relative prices p_{jt}. The problem of the firm is to choose the path of capital, employment and energy use, $\{K_t, N_{0t}, E_{1t}, E_{2t}\}_{t=0}^{\infty}$, to maximize discounted profits given by

$$\Pi_0 = \sum_{t=0}^{\infty} q_t^0 [F_(S_t, A_{0t}, N_{0t}, K_{0t}, E_{0t}) - r_t K_t - w_t N_{0t} - \sum_{j=1}^{2} p_{jt} E_{jt}] \qquad (14)$$

where q_t^0 is the price of one unit of consumption in period t in terms of consumption in period zero, and E_t is defined in (2).

A representative firm in the oil sector ($j = 1$) owns the oil stock and faces a per-unit τ_{1t} on oil extraction. R_0 is the initial stock of oil. The problem of the firm is to choose a path of oil extraction $\{R_t\}_{t=0}^{\infty}$ to maximize discounted profits given by

$$\Pi_1 = \sum_{t=0}^{\infty} q_t^0 [(p_{1t} - \tau_{1t})(R_t - R_{t+1})] \tag{15}$$

The representative firm in the sector ($j = 2$) operates the technology (4) and faces a per-unit tax equal to τ_{2t}. Naturally, a negative tax rate indicates a subsidy. The problem of the firm is to maximize discounted profits given by

$$\Pi_2 = \sum_{t=0}^{\infty} q_t^0 [(p_{2t} - \tau_{2t}) f(A_{2t}, N_{2t}, K_{2t}) - w_t N_{2t}] \tag{16}$$

There is a representative household which derives utility from the consumption of the single good in the economy and owns the firms. Households consume and work subject to the following present value budget constraint

$$\sum_{t=0}^{\infty} q_t^0 [C_t + K_{t+1}] \le \sum_{t=0}^{\infty} q_t^0 [r_t K_t + w_t N_t + T_t] + \Pi \tag{17}$$

where $\Pi = \sum_{j=0}^{2} \Pi_j$ are dividends from the firms, and T_t represents a lump sum rebate from the government. The household's problem is to choose a sequence $\{C_t, N_t\}_{t=0}^{\infty}$ to maximize (7) subject to (17), taking prices and taxes as given.

A government collects and pays subsidies. Any surplus (or deficit) is rebated in a lump sum transfer to households. The government budget constraint is given by

$$\sum_{j=1}^{2} \tau_{jt} E_{jt} = T_t \tag{18}$$

Energy Use

Considering oil firms' decisions in the market, it holds true again that the Hotelling equation governs the use of the non-renewable resource. Oil firms maximize profits when the marginal benefit of extracting oil is the same at any point in time. If the benefits are higher in the future, then it is better to postpone extraction.

At any given point in time, this benefit is equal to the proceeds from selling oil minus the tax paid to the government. This is captured in the following equation that resembles Eq. (12):

$$\beta u'(C_{t+1})[F'_{e_1,t+1} - \tau_{1,t+1}] = u'(C_t)[F'_{e_1,t} - \tau_{1,t}] \tag{19}$$

Renewable energy firms equalize marginal benefits to marginal costs. The price of renewable energy net of any tax payments determines the benefits of producing renewable energy. Costs are the cost of hiring workers. Also, economic efficiency requires that the factor input payment is their marginal productivity. Overall, the following condition represents these market trade-offs:

$$[f'_{n,t} - \tau_{2,t}]F'_{e_2,t} = F'_{n,t} \tag{20}$$

Optimal Carbon Tax

The design of optimal policies consists of finding the taxes (on carbon emissions and renewables) that induce efficient energy use in the free market. When comparing the Hotelling rule that holds in the market with the optimal rule, Eqs. (12) and (19), a problem arises to find the tax that makes these two equations match exactly.

It follows from simple observation that the carbon tax must equal the social cost of climate change for the two equations to be equal. It is the carbon tax that induces the market economy to use non-renewable resources efficiently. Thus, the optimal carbon tax rate is equal to the present value of the output losses from a marginal increase in carbon emissions.

It is a complicated concept but, under certain assumptions that restrict the social welfare function and the climate system dynamics, the carbon tax depends on three elements::

1. **Discount rate**. The cost of climate change is the discounted sum of all the damages that happen over time. Of course, if the current generation does not care about the future of next generations in fifty to one hundred years, then the optimal carbon tax is lower. The more impatient and short run oriented society is, the lower the optimal carbon tax will be.
2. **Carbon natural reabsorption**. Nature eventually reabsorbs the carbon that reaches the atmosphere, but at a slow rate. The lower this natural reabsorption is, the higher the optimal carbon tax is because the tax must fix what is not fixed by the planet itself.
3. **Climate damages**. The last critical component of the optimal tax is the magnitude of economic and social welfare damages from climate change. The more serious the damage is, the higher the tax must be. There is a substantial amount of uncertainty around this value, and there is an active and growing literature seeking to estimate these damages. It is challenging to add different damages (economic losses, health costs, biodiversity loss, among others) and measuring them in the same unit. Putting a price on carbon emission involves pricing different and diverse damages that result from the global temperature rise and resulting climate change.

It is encouraging that, despite its complicated details and various challenges, the problem of putting a price on carbon emissions comes down to the estimation of three parameters only. It is useful because it highlights the three most essential parts on which the scientific community, governments and society must focus.

Optimal Renewable Subsidies

The optimal renewable subsidy induces economic efficiency in the market economy. It follows from simple observation of Eqs. (13) and (20) that the optimal subsidy for renewable energy is zero. Notably, a carbon emissions tax is enough to induce individuals to internalize the climate change problem and act accordingly.

There is, however, worldwide proliferation of subsidies for renewable energy. Many countries seek the promotion of renewable energy as an alternative to a carbon tax. What are the arguments in favor of subsidizing renewable energy?.

There are many. One reason is the existence of externalities in the production of renewable energy. For example, a possible rationale for these subsidies is that the economy needs to learn how to produce renewables, a relatively new technology. There is a learning-by-doing process that requires initial government support. A similar case comes from the presence of productivity spillovers that individuals do not internalize. These are valid reasons to subsidize the industry.

In much the same way as the optimal carbon tax, the optimal subsidy rate should equal the externality's value. Thus, governments must be capable of measuring the externality in renewable energy generation to be able to implement optimal support. Otherwise, renewable subsidies will add undesirable distortions to the production process.

The Risks of a Green Paradox

There is a related case for promoting renewable energy. This case states that, although renewable energy is initially a more costly technology, the economy must invest in it to displace dirty fossil energy.

The puzzle is that a counterproductive effect can arise. In particular, investment in renewables may lead to an emission increase in the short term. The literature calls this effect a Green Paradox (see Sinn 2008; van der Ploeg and Withagen 2012a, b; Gerlagh 2011; Jensen et al. 2015; Belfiori 2021, among others). A Green Paradox occurs when a climate policy implemented with good intentions leads to an increase in carbon emissions rather than a reduction, at least in the short term.

Why does the Green Paradox happen? Oil companies extract oil so that the benefits of doing so are the same at any point in time. Suppose that the economy reaches the point of full phase-out to renewable energy. In that case, the benefit of leaving oil reserves for the future is zero. There is no more business moving forward. Consequently, there are incentives to extract more oil sooner than later. In fact, by solving backward, there are incentives to accelerate extraction at every point before the economy migrates to a hundred percent renewable energy.

There is a cake-eating analogy that often proves insightful. Suppose that you bought a cake, planning to eat a piece each day of the week. However, you anticipate that your partner will eat most of it within a day. Then, you better eat it faster. Unfortunately, something similar holds true for global oil reserves.

More extraction implies more pollution. Eventually, pollution will fall because production relies only on renewable energy. The problem is, however, that carbon emissions accumulate. The first increase in carbon emissions will have long-lasting climate impacts.

Final Discussion

Despite decades of climate negotiations, global emissions of greenhouse gases are still rising. The solution to the climate problem is straightforward from an economic point of view: putting a price on carbon. This price must be the same across countries and depends on three elements: discount rate, carbon depreciation rate and climate damages involved.

A carbon tax is the solution to climate change: It generates incentives both to reduce emissions and to invest in renewables. With a carbon tax, companies using fossil energy become relatively more expensive, while companies relying on alternative energy sources are relatively cheaper.

Given the unsuccessful international implementation of a global carbon tax, countries individually seek alternatives to the carbon tax, especially renewable energy subsidization. However, renewable energy subsidies are not a substitute for carbon emissions control through a carbon tax. Likewise, they can lead to an overall increase in global carbon emissions. Subsidies may still be necessary if there are externalities in the sector, typical of infant industries.

An essential takeaway of the analysis is that pursuing second-best policies is dangerous: They come at a cost to local governments and economies, while they may be unsuccessful in reaching the ultimate goal. The Green Paradox is an example.

References

Acemoglu D, Aghion P, Bursztyn L, Hemous D (2012) The environment and directed technical change. Am Econ Rev 102(1):131–66

Archer D (2005) Fate of fossil fuel CO_2 in geologic time. J Geophys Res: Oceans 110(C9):C09S05

Barrage L (2018) Optimal dynamic carbon taxes in a climate-economy model with distortionary fiscal policy. Rev Econ Stud

Belfiori ME (2021) Fossil fuel subsidies, the green paradox and the fiscal paradox. Econ Energy Environ Policy 10(1)

Gerlagh R (2011) Too much oil. CESifo Econ Stud 57(1):79–102

Golosov M, Hassler J, Krusell P, Tsyvinski A (2014) Optimal taxes on fossil fuel in general equilibrium. Econometrica 82(1):41–88. https://doi.org/10.3982/ECTA10217

Hotelling H (1931) The economics of exhaustible resources. J Polit Econ 39(2):137–175

Jensen S, Mohlin K, Pittel K, Sterner T (2015) An introduction to the green paradox: the unintended consequences of climate policies. Rev Environ Econ Policy 9(2):246–265. https://doi.org/10.1093/reep/rev010

Nordhaus WD, Boyer J (2003) Warming the world: economic models of global warming. MIT Press

Sinn H-W (2008) Public policies against global warming: a supply side approach. Int Tax Public Fin 15(4):360–394

Stern N (2008) The economics of climate change. Am Econ Rev 98(2):1–37

van der Ploeg F, Withagen C (2012a) Is there really a green paradox? J Environ Econ Manage 64(3):342–363. http://www.sciencedirect.com/science/article/pii/S0095069612000927

van der Ploeg F, Withagen C (2012b) Too much coal, too little oil. J Public Econ 96(1–2):62 – 77. http://www.sciencedirect.com/science/article/pii/S0047272711001460

WorldBank and Ecofys (2018) State and trends of carbon pricing 2018

Latin American Challenges for the Next Generation

Natural Resources, Climate Change and Governance

Carlos Winograd

Abstract As per the recent 2019 World Population Prospects, the world population will reach 9.7 billion in 2050, and this rise in population will lead to a 70% increase in the demand for agricultural production. Given the wrath of climate change and resource constraints in Asia, Latin America has a comparative advantage in terms of physical resources which make it the most plausible contender for the role of world food factory at large, considering value added incorporating biotechnology advances, logistics and the contribution of new technological development. The paper considers the local and global governance challenges due to economic and social disruptions, resulting from this significant opportunity for growth and development in the current world scenario.

Keywords Climate change · Governance · Population growth · Land use · Water use

Introduction

As per the recent 2019 World Population Prospects, the world population will reach 9.7 billion in 2050, and this rise in population will lead to a 70% increase in the demand for agricultural production. Given the wrath of climate change and resource constraints in Asia, Latin America has a comparative advantage in terms of physical resources which make it the most plausible contender for the role of world

This chapter draws partly on a conference presentation at the 7th AfD/EUDN Conference, "Fragmentation in a globalised world", hold on December 9th, 2009 in Paris, and Garcette and Winograd (2009). I thank particularly François Bourguignon, Luis Miotti, Marcos Orteu, Robert Pecoud, Thierry Verdier and Alejandro Winograd for valuable comments. I am also grateful to participants in seminars in Paris, London, Montevideo, Buenos Aires and Rio de Janeiro. This chapter benefited significantly from the assistance and discussions with Resuf Ahmed. I thank particularly Elisa Belfiori for her comments and the encouragement for the publication in this book.

C. Winograd (✉)
Paris School of Economics and University of Paris-Evry Val d'Essone, Paris, France
e-mail: winograd@pse.ens.fr

© Springer Nature Switzerland AG 2021
M. E. Belfiori and M. J. Rabassa, (eds.) *The Economics of Climate Change in Argentina*, The Latin American Studies Book Series,
https://doi.org/10.1007/978-3-030-62252-7_7

food factory at large, considering value added incorporating biotechnology advances, logistics and the contribution of new technological development.

This chapter discusses the prospects of a development path for the region capturing the benefits of the potential response to world food demand. We will briefly introduce and discuss the trade-offs between the productive capacity to engage in the above strategy and the environment and social tensions arising from such a promising road for economic growth and development. The local and global governance challenges resulting from this development model in the current world scenario will also be considered. The chapter includes four sections. After this introduction, the second section presents the basics of world population, the supply and demand forecasts for food, as well as the physical endowments of Latin America in land and water; the third section discusses the trade-offs between the production scenarios and climate change, sustainability and governance tensions. The fourth section contains the conclusions.

Natural Resources and Neo-Malthusianism

Population and Food

Latin America accounts for about 8.4% of the global population today, and it is expected to account for 7.8% by 2050 and 6.3% by 2100 (Table 1).[1] The region has almost achieved its demographic transition with total fertility of 2.0 children per woman on average in 2019, against 5.83 in 1960s. This transition was, in many cases, achieved with a chaotic family planning, or even with the absence of it. Brazil, for example, did not have any active plan but fertility habits have changed dramatically. Today, the fertility rate is less than two children per woman, and the demographic growth is very close to European standards.

In the next decades, Asia and Africa are expected to experience the relatively highest rates of growth in population, and the former is expected to show a significant rise in income, whereas this is more uncertain for the latter. Concerning the world population, there will be almost two billion Asians and two billion Africans more by the year 2050, which adds up to four billion people more on the planet. Asia, in particular, is likely to experience strong income growth and will, therefore, combine two main drivers of world consumption.

Are we in face of neo-malthusian era, in the framework of growing middle classes? In particular, with more people and a higher income, Asians will increase the demand for food products (richer in protein and more diverse)[2] in the world. In the case of Africa, income growth might also come with the population growth, but this is more uncertain despite a significant stock of natural resources, due to severe local governance problems. Political instability, ethnic and religious conflict, weak institutions,

---

[1] World Population Prospects 2019, Department of Economic and Social Affairs, United Nations.
[2] FAO, Regional Office for Asia and the Pacific (2008) and Alexandratos and Bruinsma (2012).

Table 1 Population and fertility rates, world and SDG regions (2019, 2050, 2100)

Region	Population (millions)				Average number of live births per woman			
	1960	2019	2050	2100	1960	2019	2050	2100
World	*3034*	*7713*	*9735*	*10,875*	*5.02*	*2.5*	*2.2*	*1.9*
Sub-Saharan Africa	220	1066	2118	3775	6.64	4.6	3.1	2.1
Northern Africa and Western Asia	129	517	754	924	6.55	2.9	2.2	1.9
Central and Southern Asia	619	1991	2496	2334	6.05	2.4	1.9	1.7
Eastern and South Eastern Asia	1019	2335	2411	1967	5.62	1.8	1.8	1.8
Latin America and the Caribbean	220	648	762	680	5.83	2.0	1.7	1.7
Australia/New Zealand	12	30	38	49	3.38	1.8	1.7	1.7
Oceania	3	12	19	26	6.31	3.4	2.6	2.0
Europe and Norther America	810	1114	1136	1120	2.74	1.7	1.7	1.8

Source by author, based on data of UN (2019)

Earth's Surface	Land: 29% (149 M km²)			Ocean: 71% (361 M km²)	
Land Surface	Habitable Land: 71% (104 M km²)		Glaciers: 10% (15 M km²)	Barren Land: 19% (28 M km²)	
Habitable Land	Agriculture: 50% (51 M km²)	Forrests: 37% (39 M km²)	Shrub: 11% (12 M km²)	Urban and built-up land: 1% (1,5 M km²)	Freshwater: 1% (1,5 M km²)
Agricultural Land	Livestock (growing land and feed production): 77% (40 M km²)			Crops: 23% (11 M km²)	

Fig. 1 World land distribution and uses. *Source* By author, based on Our World in Data

including very fragile judicial systems imply high discount rates for investment projects jeopardizing growth prospects. Latin America and North America are also expected to show some population growth, while population growth in Europe is in decline.

In the case of Europe, the role of immigration in its demographic dynamics and the sustainability of immigration over time is still an open question. In any case, the growth in world population and world income leads to a growing demand for food. According to estimates by the Food and Agriculture Organization, there will be a 70% increase in the world demand for agricultural products by 2050. Will Latin America be able to increase its agricultural production to meet some of these future needs? An exciting possibility arises here. Historically, the volatility in commodity prices has been a challenge for countries in which productions depend heavily on agricultural products. However, if the market for agricultural products shows persistent dynamism, prudent public policies may mitigate the intrinsic price volatility.

Latin America might find out that it has a new comparative advantage in the agri branch at large and will have to seek the appropriate means to capitalize on this advantage. The established view of inevitable decrease in the terms of trade maybe in question beyond the recent phase of sharp rise and relative decline in prices.[3] Latam, and particularly South America may face a "neo-malthusian" global era, reshaped in view of climate change and new dimensions of scarcity. If a proper balance in product specialization should be aimed by policy-makers, the neglect of these trends in long-term dynamics may lead to miss significant opportunities for economic and social development.

Cultivated Land

Of the world habitable land (104 million km²), 50% is dedicated to agriculture, some 51 million km² (. 1), of which 23% is crops for food, and 77% allocated to livestock (this includes land to produce feed for livestock). Land is a finite resource, in a world of growing population (Ritchie and Roser 2013).

---

[3] Since the early years of the twenty-first century, we have observed a sustained rise in international markets food commodity prices. If the boom years seem behind, despite the big global recessions, of 2008 with the financial crisis, and of 2018 with the Covid 19 pandemia, international market prices stay high as compared to the levels experienced since the 1990s and before (FAO 2020). Forecasts for 2050 and beyond point to the sustainability of this path (FAO 2018).

Asia faces a strict limit on the extension of its agriculture land. As mentioned above, this continent is the most significant contributor to world demand because of its population and income growth, but it does not have the necessary land to meet its demand. This explains China's attempts to buy land abroad, engineered diverse agreements to secure resources in Africa and Latin America: Asians face a growing demand but do not have the extension and quality of land to produce its consumption needs. In the past, colonial wars were the solution to this dilemma. But in the future, the world will have to find better means to solve it.

Where is cultivable land? Latin America accounts for 24% of potentially world arable land (Table 2) and 17% of the pastures (FAO 2011). Therefore, Latin America has a good stock of physical (natural resource) capital needed to support the increase in world production. Latin America has only 9% of the world's cultivated land in 2020, and the region can nearly double its cultivated land by 2050 (from 142 to 228 million ha, Table 3), given the starting point and the current productive conditions.[4] Additionally, more than 90% of Latin America land cover is cultivated, forest or grasslands (Fig. 2).[5]

According to various sources of scenarios and forecasts, Latam may increase its production by 50 to 100% between 2020 and 2050, or double to triple since 2000.[6] This sharp increase in supply potential is the result of a significant increase in cultivated land coupled with varying degrees of productivity growth. We should highlight a yield gap of about 52%, which means that in Latin America actual yield is just 48% as compared to the potential yield. In contrast, the yield gap in East Asia is just 11%, showing the constraint in food production capacity of the latter region with very limited land resources (Table 4). Latin America can take advantage of this opportunity.[7] On the demand side for agricultural production, the biodiesel branch sharp increase in production adds to food leading to higher pressure on world agricultural supply and prices (Fisher 2009), in ranges from 10 to 60% depending on the share of first generation biofuels in transport fuels (1–8%).[8] But even before considering technological improvements, arises the question on the opportunity for Latam to become the "world food powerhouse" of the twenty-first century? An open question in motion.

Additionally, meat demand and production will increase globally. Two effects drive the additional demand, the increase in world population and higher incomes in developing nations such as South and East Asia.[9] It is estimated that global meat production will have to increase from 258 million tons globally in 2007 to 455 million

---

[4]FAO 2011. The state of the world's land and water resources for food and agriculture.

[5]FAO 2011, Page 22 (adapted from Fischer et al. (2011)). See Fig. chap 1.1 in the Appendix.

[6]See Agrimonde (2009), Agrimonde et al. (2014), Allan et al. (2019), Fisher (2009), Alexandratos and Bruinsma (2012).

[7]FAO 2011, Page 37 ((adapted from Fischer et al. (2011)). See Table chap 1.7 in the Appendix.

[8]See Fisher (2009).

[9]For example, Alexandratos and Bruissma (2012) estimate that the daily energy supply (calories per capita per day) will increase from 2772 in 2007 to 3070 in 2050.

Table 2 Land with rainfed crop production potential (million ha)

Region	Total land surface	Suitable land (1) Total (A + B)	Prime (A)	Good (B)	Not usable (2)	Suitable for use (3 = 1-2)
World	13,295	4495	1315	3180	1824	2671
Developing countries	7487	2893	816	2077	1227	1666
Sub-Saharan Africa	2281	1073	287	787	438	635
Latin America	2022	1095	307	788	580	515
Near East/North Africa	1159	95	9	86	9	86
South Asia	411	195	78	117	43	152
East Asia	1544	410	126	283	140	270
Other developing countries	70	25	9	15	23	2
Developed countries	5486	1592	496	1095	590	1002
Rest of the world	322	11	3	8	7	4

Source by author, based on data from GAEZ v3.0

Table 3 Land in use, 2009–2050, forecast by region. Cultivated (C) and harvested (H) in million hectares and cropping share of world total (CSW) and cropping intensities (CI) in %

Region	2009				2050			
	C (M ha)	CSW (%)	H (M ha)	CI (%)	C (M ha)	CSW (%)	H (M ha)	CI (%)
Asia	542	35	588	109	541	32	641	118
South Asia	204	13	232	113	212	13	243	115
East Asia	133	9	176	133	133	8	191	144
Southeast Asia	101	7	111	109	107	6	124	115
Rest of Asia	103	7	70	68	88	5	83	94
Americas	395	26	279	59	468	28	384	82
Northern America	253	17	146	58	241	14	192	80
Latin America	142	9	128	90	228	14	193	85
Africa	251	16	214	85	342	20	270	79
Northern Africa	28	2	21	74	27	2	25	92
Sub-Saharan Africa	223	15	194	87	315	19	245	78
Europe	293	19	184	63	264	16	219	83
Oceania	46	3	26	57	58	3	48	83
World	**1527**	**100**	**1286**	**84**	**1673**	**100**	**1562**	**93**

Note Difference between Total Land Use (1) and Rainfed Use (2), is Irrigated Use (3), so 1 = 2+3. Crop intensity is the ratio between Harvested and Cultivated Land, so H/C = CI

Source by the author, based on date of FAO (2011)

Definition The fraction of the cultivated area that is harvested. The cropping intensity may exceed 100 percent where more than one crop cycle is permitted each year on the same area. In AQUASTAT, the cropping intensity has been calculated on irrigated crops only and becomes practically the ratio of the harvested irrigated areas over the area equipped for full control irrigation actually irrigated. Irrigation, by decoupling the crop production from the natural precipitation, increases cropping intensity in countries where temperatures are not a limiting factor. Source: http://www.fao.org/nr/water/aquastat/data/glossary/search.html?termId=7587&submitBtn=s&cls=yes

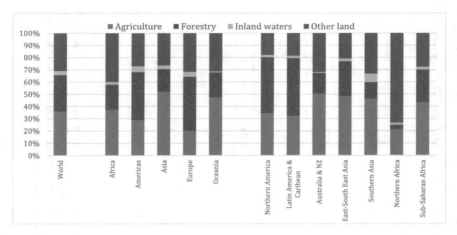

Fig. 2 Regional distribution for land use and cover (agriculture area includes arabal land, permanent crop and land for grazing. Forestry includes primary forrest, regenerated forrest and planted forrest. Other land includes barren land and urban/built-up land)uthor, based on data. *Source* by author, based on data of FAO STAT

Table 4 Ranking of yield gap by region and percentage of potential

Region	Yield gap 2005 (%)
Sub-Saharan Africa	76
Eastern Europe and Russia	63
Northern Africa	60
South Asia	55
Southern America	52
Australia and New Zealand	40
Western and Central Europe	36
Norther America	33
Southeast Asia	32
East Asia	11

Source by the author, based on data of FAO (2011)

tons in 2050, a 70% rise.[10] Most of this increase in meat production will take place in developing countries that may double their supply in the next 40 years, and Latin America will have an important role to play (Tables 5 and 6).

Latin America, however, shows a handicap in "logistics" at large.[11] The region performs relatively low in terms of competitive infrastructure to reduce the constraints to land accessibility. Nelson (2008) constructed a data set of "travel time to major cities" to quantify accessibility of land resources, showing relevant limitations from this perspective. For currently cultivated land, only 32% of cultivated land is accessible within 2 h as compared to 85, 62 and 53% in Western Europe, Northern Africa and South Asia, respectively. For accessibility of prime and good unprotected grassland/woodland ecosystems and unprotected prime and good forestland ecosystems, the situation is much worse in Latin America, where the same figure stands at 23% and a meager 7%, respectively. For Latin America to become the breadbasket of the world, investment in agricultural infrastructure will play a pivotal role.[12]

World Water Distribution

For the "world food powerhouse" land is necessary, but not enough. Water is crucial, and Latin America has one-third of the world permanently renewable freshwater. Thus, Latin America not only has the land but also has the *fuel* to support the land: water. Every here and there, conspiracy theories emerge on old and new imperialisms campaigning to seize Latam water. And while this might be an unfounded conspiracy fear, there is a tied bond between the increase in world demand and the world supply of water. Latin America is the one place in the world where there is land, and there is also an excess supply of renewable water, being the largest reserve worldwide (Table 7).

In contrast to Latam water availability, there are already very severe water shortages, most particularly in Western, Central and South Asia, that demand more then half of their water resources in irrigation (Table 7). In Northern Africa, withdrawals for irrigation exceed renewable resources due to groundwater overdraft and recycling. On the other hand, South America barely uses 1% of its resources.[13]

---

[10]The long-term view of meat production (all types) estimates from Alexandratos and Bruissma (2012).

Meat production (million tons)	1961/1963	2005/2007	2050
World	72	258	456
Developed	52	109	137
Developing	20	149	317

[11]See Schwab and Sala-i-Martín (2016).

[12]See Table 11, 12, 13 in Fischer et al. (2012).

[13]FAO (2011). The state of the world's land and water resources for food and agriculture (SOLAW).

Table 5 Meat (all types) aggregate production (2007, 2018) and forecast (2050)

Region	2005/2007	2018	2050*	1971–2007	1991–2007	2007–2050	2005/2007	2050
(Unit)	('000 tons)			(growth rate, %)			(% of total)	
Sub-Saharan Africa	6802	11,557	23,928	2.4	3.0	2.9	2.6	5.2
Near East/North Africa	8918	14,072	23,233	4.7	3.9	2.2	3.5	5.1
Lat. Amer. and Carib.	40,585	56,143	71,644	3.9	4.5	1.3	15.7	15.7
South Asia	7180	12,645	40,327	2.8	1.6	4.0	2.8	8.8
East Asia	85,121	112,961	156,930	6.6	4.5	1.4	32.9	34.4
Developed countries	109,424	124,282	136,276	1.1	0.6	0.5	42.4	29.9
World	258,370	342,396	456,097	2.8	2.4	1.3	100	100

Source by author, based on Alexandratos and Bruinsma (2012) and FAO STAT

Table 6 Meat (all types) aggregate consumption (2007), forecast (2050) and structure

Region	2005/2007	2050*	1971–2007	1991–2007	2007–2050	2005/2007	2050
(Unit)	('000 tons)		(growth rate, %)			(% of total)	
Sub-Saharan Africa	7334	26,926	2.4	3.0	2.9	2.9	6.0
Near East/North Africa	10,292	27,992	4.7	3.9	2.2	4.0	6.2
Lat. Amer. and Carib.	34,557	61,003	3.9	4.5	1.3	13.5	13.5
South Asia	6685	40,858	2.8	1.6	4.0	2.6	9.0
East Asia	86,806	160,037	6.6	4.5	1.4	33.9	35.4
Developed countries	109,382	130,385	1.1	0.6	0.5	42.7	28.8
World	256,179	452,230	2.8	2.4	1.3	100	100

Source by author, based on Alexandratos and Bruinsma (2012) and FAO STAT

Table 7 Renewable water resources and irrigation (annual average)

Continent regions	Precipitations (mm)	Renewable water (RW) resources (km^3)	Percentage of world RW total (%)	Water use efficiency ratio (%)	Irrigation water withdrawal (km^3)
World	**809**	**43,022**	**100**	**44**	**2710**
Africa	678	3931	9	48	184
Northern Africa	96	47	0	69	80
Sub-Saharan Africa	815	3884	9	30	105
Americas	1091	19,238	45	41	385
Northern Americas	636	6077	14	46	258
Central America and Caribbean	2011	781	2	30	15
South America	1604	12,380	29	28	112
Asia	827	12,413	29	45	2012
Western Asia	217	484	1	47	227
Central Asia	273	263	1	48	150
South Asia	1602	1766	4	55	914
East Asia	634	3410	8	37	434
Southeast Asia	2400	6490	15	19	287
Europe	540	6548	15	48	109
Western and Central Europe	811	2098	5	43	75
Eastern Europe and Russia	467	4449	10	67	35
Oceania	586	892	2	41	19
Australia and New Zealand	574	819	2	41	19
Pacific Islands	2062	73	0	–	0.05

Source by author, based on FAO (2011)

Therefore, Latin America can respond to a sharp increase in world demand for food and agri products at large. It has resources and tools. No other region in the world has the means to do it, except probably Sub-Saharan Africa. The problem with Africa is that although the continent has the quality of land and renewable water to participate in the food supply expansion strategy, as mentioned earlier, there is a severe and widespread governance problem coupled with weak property rights, which tends to distort appropriate incentives and leads to extremely high discount rates and thus impedes the necessary capital investments.

On the other hand, it is in Africa where the risk of hunger will remain high in the face of stress on the food markets worldwide. In 1970, there were 950 million people in risk of hunger, 26% of a total world population of 3.7 billion inhabitants. In 2020, the total population in risk of hunger is estimated to be near 830 million, 11% of 7.8 billion inhabitants, whereas the forecasts for 2050 are 460 million, 4.7% of a total population of 9735 billion inhabitants. In the face of a sharp reduction in absolute and relative population in risk of hunger, Sub-Saharan Africa accounted for 10% in 1970, whereas it will represent more than 50% in 2050 (with rather stable absolute numbers since 2010), Asia 35% and Latin America 2%.[14]

Natural Resources, Climate Change and Conflicts

In a nutshell, increases in world population and income growth give to Latin America the demand counterpart to expand production through agriculture. This is also true for metals, where the region has again enormous resources to supply world demand. But, while feasible, the "world granary strategy" has an underlying problem. An increase in the cultivated area requires deforestation. In turn, deforestation leads to climate change and soil pollution. Furthermore, an increase in the cultivated area may imply population displacements in many countries and thus requires the appropriation of aboriginal land. Higher wealth and income inequality, as well as, ethnic and social conflicts may result.

All such trends of economic and social development may show in the future but are already in the scene today. It is what happens in the Amazon these days. Landowners in the south of Brazil are moving to much cheaper land in the Mato Grosso. They can sell an acre in the south and acquire much bigger properties becoming landowners in the north. But, on the way to this agricultural transformation entire aboriginal communities and subsistence farmers are pushed out of their lands, and massive deforestation is needed to prepare the land for production.[15]

---

[14]Based on UN (2019) and IASA world food system simulations, scenario FAO-REF-00 (May 2009), see also Fisher (2009). The population in risk of hunger in Sub-Saharan Africa was 85 million in 1970, 286 million in 2020 and estimated at 240 million in 2050, whereas in East Asia falls from 500 million in 1970 to 26 million in 2050.

[15]Fonseca et al. (2020). Boletim do desmatamento da Amazônia Legal (maio 2020) SAD (p. 1). Belém: Imazon. https://imazon.org.br/publicacoes/boletim-do-desmatamento-da-amazonia-legal-maio-2020-sad/.

Climate Change and Governance

The natural ressource-based strategy highlights two dimensions of conflict: on the one hand, the social conflict related to the extension of the agricultural frontier and the impact on migration, ethnic tensions, as well as potential changes on wealth and income inequality. On the other hand, this development path stresses climate change dynamics. These two problems lead to challenges in local and global governance, respectively. Local governance refers to the economic and social tensions within countries, whereas global governance arises from international tensions between countries related to this development compact. Beyond classical international trade conflicts and protectionist policies, global governance adds today climate change as a new key driver. The expansion of the agricultural frontier may provoke massive deforestation leading to significant global negative externalities. One wonders for how long Brazil, and its neighbor countries will be able to govern without international opinion its Amazon territories, considering that it affects the provision and quality of air on the planet. Will one day countries in the world tell Brazil that the Amazon rainforest *is too vital* for the planet to let Brazil decide on its own the destiny of territories that may influence the planet path?[16] Will the moral argument that the developed countries had freely chosen climate change dynamics of the world for two hundred years deter this trend of facts? May or will the ignorance and environment neglect of the past justify the unrestricted sovereignty argument of today? National sovereignty, international conflict and politics of cooperation thus enter the scene.

About 50% of Latin America is covered with forest. Latin America has 22% of the world's forests and 52% of the tropical forest—mainly the Amazon rainforest. Latin America is thus a crucial player in the climate change arena. It can become an active part of the problem if deforestation continues. Or it can play part of the solution if the region contributes actively to keep and preserve the 50% of the world forest that it governs today.[17] According to a variety of projections, without policy interventions, more than one-half of the Amazon forest might be destroyed or degraded by 2030–50 (Swan et al. 2015).

On the other hand, vicious circle dynamics may arise for Latam. Climate change may seriously jeopardize the development path strongly based on the "food factory strategy". If South America is not by far the sole driving force of climate change, this phenomena may in turn through higher frequency of extreme weather events, as well as changes in precipitation patterns, seriously affect hydroelectric supplies and water supply. Consequently, the stress on agriculture may thus endanger the "world granary and food" intensive engine of growth (Fisher 2009).

However, since 2008 we observe that deforestation has slowed down, and hopefully, Brazil has entered a new downward trend. Annual deforestation in Brazil's Amazon forest was lowest in 2012 since 1988. But the Brazilian government now in place headed by President Jair Bolsonaro passionately enrolled in a militant neglect

---

[16]Regular international polls show the growing concern on climate change. See Pew Research Center, Global Attitude Survey (2019).

[17]Latin American and Caribbean Forestry Commission (LACFC, FAO).

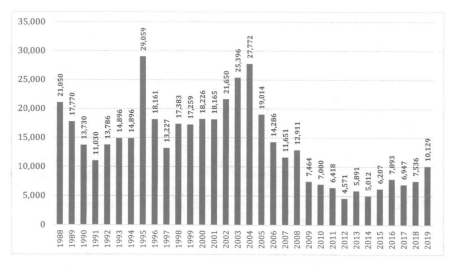

Fig. 3 Deforestation rate of the Brazilian Amazon, annual, km^2, 1988–2019. *Source* by author, based on TerraBrasilis, INPE

of the risks due to climate change in line with the position on this topic of the current US President Donald Trump. By accident or policy voluntary neglect, a very high level of deforestation has been observed in 2019, which is the highest in 11 years (Fig. 3). The coming years will show if we are in face of an outlying year or an upward change in the trend of deforestation. If Brazil follows the path initiated last year, the country will be unlikely meet its climate commitments made as a participant of the Paris Agreement (2015). Ongoing deforestation in the Brazilian Amazon can have long-term implications which can further fuel, sooner or later, severe local and global governance conflicts.

Conclusions

This chapter proposes a discussion on the potential of Latin America, particularly South America, to leverage on its natural resources to benefit from a "neo-Malthusian" scenario in the global economy for the next generation. We focus on the world market for food at large and discuss the potential demand and supply looking to 2050.

Highlighting the world food supply advantage should not lead the reader to neglect other significant opportunities for economic and social development in the region. One should consider dynamic knowledge services (TICS), tourism, certain manufacturing sectors or other natural resources industries.

We show that the demand will have a significant increase driven by a strong growth in population in Asia and Africa coupled with a sharp rise in income in the former, while rather uncertain in the latter. Biofuels used in transport will add to food demand sustaining agricultural prices. Latin America has the appropriate land and renewable water, adding productivity gains, to increase production and supply the future demand for food worldwide. Asia does not have suitable land and the necessary renewable water to expand significantly production, whereas Sub-Saharan Africa could contribute with a sizeable increase in supply. However, despite abundant appropriate land and renewable water, the potential rise in production of the latter is extremely uncertain due to severe political and ethnic conflict, weak property rights, leading to highly disruptive discount rates that jeopardize a robust investment path for a sustained expansion in the regional supply of agriculture products. Beyond the production front, if the world population in risk of hunger will strongly decrease as a share of total population, it will be concentrated in Africa.

The Sub-Saharan African question is certainly a matter for future research. How will these scenarios and prognosis change if a strong drive for reforms, leading to lower discount rates, takes hold in the region? In the 1950s, South Korea and Ghana had the same income per capita, but economic and institutional change turned into a structural break for development in South East Asia. Could something of the sort happen in Sub-Saharan Africa in coming decades? A robust and sustained development path in the region could drive an increase agriculture production, as well as in incomes and food consumption, implying an interesting review and further debate on the topics here discussed.

In this article last section, we discuss the potential downside of the world food factory strategy, due to deforestation and climate change. On the one hand, the expansion of the agricultural frontier may lead to population displacement, social conflict and local governance tensions, while it may also lead to weather and soil quality disturbances jeopardizing the sustainability of the agricultural-based strategy. On the other hand, the growing international concern on climate change may lead Latin American countries to confront global governance tensions. Thus, policy-making in the region, to benefit from the food factory opportunity in the coming decades, requires a balanced political economy approach to manage multiple constraints.

References

Agrimonde (2009), Scenarios and challenges for feeding the world in 2050, INRA and CIRAD
Agrimonde, Paillard S, Treyer S, Dorin B (eds) (2014) Scenarios and challenges for feeding the world in 2050. Editions Quae, Springer, Berlin
Alexandratos N, Bruinsma J (2012) World agriculture towards 2030/2050: the 2012 revision. ESA Working paper No. 12–03. Rome, FAO
Allan T, Bromwich B, Keulertz M (eds) (2019) Oxford handbook of food, water and society
FAO Regional Office for Asia and the Pacific, (2008) Asia Pacific food update. Economic and Social Department Group (RAPE), Bangkok
FAO (2018) The future of food and agriculture—Alternative pathways to 2050. Rome. 224 pp
FAO (2011) The state of the world's land and water resources for food and agriculture (SOLAW)— Managing systems at risk. Food and Agriculture Organization of the United Nations, Rome and Earthscan, London

Fischer G (2009) How do climate change and bioenergy alter the long-term outlook for food, agriculture and resource availability. In: Expert meeting on how to feed the world in (vol 2050)

Fischer G, van Velthuizen H, Nachtergaele F (2011) GAEZ v3.0—Global Agro-ecological Zones Model documentation. (mimeo), IIASA, Luxemburg

Fischer G, Hizsnyik E, Prieler S, van Velthuizen H, Wiberg D (2012) Scarcity and abundance of land resources: competing uses and the shrinking land resource base. FAO, SOLAW Background Thematic Report—TR02

Fonseca A, Cardoso D, Ribeiro J, Ferreira R, Kirchhoff F, Amorim L, Monteiro A, Santos B, Ferreira B, Pontes M, Souza C Jr, Veríssimo A (2020) Boletim do desmatamento da Amazônia Legal SAD. Imazon, Belém, p 1

Garcette N, Winograd C (2009) Fragmentation, globalization and Latin America: From the financial crisis to the next generation, Mimeo, EUDN, AFD conference, Paris

Nelson A (2008) Travel time to major cities: A global map of accessibility. European Commission, Ispra

Ritchie H, Roser M (2013) Land Use. Published online at OurWorldInData.org. Retrieved from: 'https://ourworldindata.org/land-use' Agrimonde, 2009, Scenarios and challenges for feeding the world in 2050, INRA and CIRAD

Schwab K, Sala-i-Martín X (2016, April) The global competitiveness report 2013–2014: Full data edition. World Economic Forum

Swann AL, Longo M, Knox RG, Lee E, Moorcroft PR (2015) Future deforestation in the Amazon and consequences for South American climate. Agric For Meteorol 214:12–24

United Nations, Department of Economic and Social Affairs, Population Division (2019) World population prospects

Index